How Scientific Instruments Have Changed Hands

History of Science and Medicine Library

VOLUME 56

Scientific Instruments and Collections

STUDIES PUBLISHED UNDER THE AUSPICES OF THE SCIENTIFIC INSTRUMENT COMMISSION

VOLUME 5

The titles published in this series are listed at *brill.com/hsml*

How Scientific Instruments Have Changed Hands

Edited by

A. D. Morrison-Low, Sara J. Schechner & Paolo Brenni

BRILL

LEIDEN | BOSTON

Cover illustration: The Kohlrausch universal bridge in James W. Queen & Co., *Catalogue and Price-List Electrical Testing Apparatus*, Philadelphia, 1887, p. 48, detail.

Library of Congress Cataloging-in-Publication Data

Names: Morrison-Low, A. D., editor. | Schechner, Sara, 1957– editor. | Brenni, Paolo, editor.
Title: How scientific instruments have changed hands / edited by A. D. Morrison-Low, Sara J. Schechner & Paolo Brenni.
Description: Leiden ; Boston : Brill, [2016] | Series: History of science and medicine library ; volume 56 | Series: Scientific instruments and collections ; volume 5 | Includes bibliographical references and index.
Identifiers: LCCN 2016033199 (print) | LCCN 2016036426 (ebook) | ISBN 9789004324923 (hardback : alk. paper) | ISBN 9789004324930 (E-book)
Subjects: LCSH: Scientific apparatus and instruments—History—18th century. | Scientific apparatus and instruments—History—19th century. | Scientific apparatus and instruments—History—20th century. | Scientific apparatus and instruments—Marketing—History.
Classification: LCC Q185 .H75 2016 (print) | LCC Q185 (ebook) | DDC 502.8/4—dc23
LC record available at https://lccn.loc.gov/2016033199

Typeface for the Latin, Greek, and Cyrillic scripts: “Brill”. See and download: brill.com/brill-typeface.

ISSN 1872-0684
ISBN 978-90-04-32492-3 (hardback)
ISBN 978-90-04-32493-0 (e-book)

Contents

Preface

In recent years, instrument historians have begun to map out how instruments were made and by whom; who were makers and manufacturers, who were retailers and suppliers, and how their products reached their markets. The picture may not yet be complete, but it has been acknowledged that it is more complex than sketched out fifty or even twenty-five years ago. This volume—from a variety of papers produced over the past few years at the Scientific Instrument Commission's annual symposia—looks at the movement of instruments to that market place, and then beyond. The core group of papers presented here were amongst those delivered at a special symposium, *How Instruments Change Hands*, organised by Sara J. Schechner and A. D. Morrison-Low on behalf of the Scientific Instrument Commission at the XXIII International Congress of History of Science and Technology held in Budapest in July 2009.

This selection looks across time and place, for the geographical location of instruments proves to be of critical interest. Alexi Baker examines early market forces in what was then the centre of the Western instrument-making world, London, through the surviving documents of an apparently major retailer, George Willdey, whose business flourished for forty-two years. Moving forward in time, Joshua Nall and Liba Taub consider British trade literature after 1800, based on the survival of two centuries' worth of material held by the Whipple Museum of the History of Science, Cambridge, while Morrison-Low scrutinises some very particular mid-nineteenth century examples of instrument advertising. Paolo Brenni investigates German, French and London prices of specific instruments used in physics across the Western world between about 1850 and 1910. Peggy Kidwell looks at the latter part of the nineteenth century's international exhibitions, and what happened to mathematical instruments there and afterwards. Taking a transatlantic standpoint, Gloria Clifton appraises how the links between the trades on either side of the Pond were connected: through the migration of personnel, through trade literature and advertising. Schechner looks at a particular instrument—the pocket sundial—from its earliest appearance on North American soil: how it arrived there, and what it can tell us about the first owners. Staying in North America, Richard L. Kremer analyses the pricing of instruments before 1800, especially in the United States, while Laura Cházaro looks again across the Atlantic to the relationship between Europe and Mexico through nineteenth-century trade in medical instrumentation.

These papers show that new sources provide interesting insights into their subjects. The rise of the accessibility of online material has enabled modern researchers to find more detail and often new case studies to uncover the past. As well as using conventional archives in the Guildhall Library, Baker has used online Old Bailey court records and contemporary newspapers; Nall and Taub have consulted a large archive of trade literature, now made available online. Morrison-Low accessed newspaper archives consulted through the National Library of Scotland's website; Brenni looked at online sources for trade catalogues, making the point that these are indispensable tools for research into the economics of the trade. Clifton has been compiling a database, building on and supplementing her 1995 publication; Schechner has been able to access instrument collections and archaeological sites for comparative material, as well as making extensive use of digitised collections of early maps in order to determine the locations of forts and trading posts, and Kremer also used online early American newspaper archives. All these recently-available literary and visual sources often supply information that the instruments themselves cannot: yet, scrutinising the instruments—as Schechner in particular shows here—enables the scholar to pull together intriguing and useful pictures of the past. All the papers presented here are—of course—based on the primary source of the artefacts themselves. Although this volume illuminates only some aspects of the scientific instrument trade, we hope that nevertheless they reveal the complexities and the multifaceted nature of this intriguing subject.

Here is a collection of case-studies from the United Kingdom, the Americas and Europe showing instruments moving from maker to market-place, and, to some extent, what happened next.

A. D. Morrison-Low
Sara J. Schechner
Paolo Brenni

February 2016

List of Illustrations

Colour Plates

Figures

Tables

Notes on Contributors

Alexi Baker

Dr, is a Mellon-Newton Interdisciplinary Research Fellow at the Centre for Research in the Arts, Social Sciences, and Humanities at the University of Cambridge. Dr Baker is affiliated to the Department of History and Philosophy of Science, at which she previously worked on Simon Schaffer's AHRC-funded project on the British Board of Longitude and on the digitization of the Board archives. She has been researching the instrument trade of early modern London and its products since 2004.

Paolo Brenni

graduated in experimental physics in 1981 from the University of Zürich. He then specialized in the history of scientific instruments and the precision industry from the beginning of the eighteenth century to the mid-twentieth century. He is a researcher for the Italian National Research Council (CNR), works in Florence for the Fondazione Scienza e Tecnica, and collaborates with the Museo Galileo. He has catalogued, reorganized, and restored several collections of instruments both in Italy and aboard. He has also written several articles dealing with instrument history, restoration, and the development of instrument industry and trade. He is currently president of the Scientific Instrument Society.

Laura Cházaro

is a professor and researcher at the Centro de Investigaciones y de Estudios Avanzados del Instituto Politécnico Nacional in Mexico. She is currently studying notions of objectivity and the uses of medical and mathematical instruments in eighteenth- and nineteenth-century Mexico. In 2007, she published "La cultura médica instrumental: los viajes políticos de los esfigmógrafos, entre Europa y América" in the journal *Nuevo Mundo, Mundos Nuevos*.

Gloria Clifton

PhD, until her recent retirement, was the curatorial head of the Royal Observatory, Greenwich, and curator of scientific instruments at the National Maritime Museum, both institutions being part of the Royal Museums Greenwich in London, UK. She is compiler of the *Directory of British Scientific Instrument Makers, c. 1550–1851*, published in 1995, and is currently working on a sequel on British instrument makers, 1851–1914.

Peggy Aldrich Kidwell
a historian of science by training, is Curator of Mathematics at the Smithsonian's National Museum of American History. She has published on historical aspects of astrophysics, mathematical instruments, mathematics education, and computing. Topics of her exhibitions range from arithmetic teaching to computer bugs to post-war American science. Her present research concerns the history of mathematical recreations. She is now working with others to display Smithsonian mathematical instruments in an online encyclopaedia.

Richard L. Kremer
teaches history of science at Dartmouth College and curates that university's collection of historic scientific instruments. Each year he teaches a seminar on the material culture of science in which his students create an exhibit from that collection. With David Pantalony and Francis Manasek he authored *Study, Measure, Experiment: Stories of Scientific Instruments at Dartmouth College* (Norwich, VT, 2005). He has also published numerous articles on nineteenth-century physics pedagogy, high-speed cinematography, and early modern astronomy and its material culture.

A. D. Morrison-Low
won the 2008 Paul Bunge Prize for her book *Making Scientific Instruments in the Industrial Revolution* (Aldershot, 2007). She has been a member of the Editorial Board for the Scientific Instrument Commission's series *Scientific Instruments and Collections* since 2009, and she is now a Research Associate, at National Museums Scotland.

Joshua Nall
is Assistant Curator of the Whipple Museum of the History of Science, in the Department of the History and Philosophy of Science at the University of Cambridge. His research focuses on mass media and material culture of the physical sciences after 1800. He is currently preparing a monograph on the role of mass media in *fin de siècle* debates over life on Mars.

Sara J. Schechner
is the David P. Wheatland Curator of the Collection of Historical Scientific Instruments and Lecturer on History of Science at Harvard University. She serves on the Editorial Board for the Scientific Instrument Commission's series *Scientific Instruments and Collections.* Her recent books include *Time and Time Again: How Science and Culture Shape the Past, Present, and Future* (Cambridge,

MA, 2014) and *Tangible Things: Making History through Objects* (Oxford, 2015), co-authored with Laurel Thatcher Ulrich, Ivan Gaskell, and Sarah Carter. Dr Schechner's research has focused on time-finding and social change; instruments of glass; colonial astronomy; and workshop practices in America.

Liba Taub
is Director and Curator of the Whipple Museum of the History of Science, and Professor of History and Philosophy of Science, at the University of Cambridge. Her research focuses on material culture of science, and Greco-Roman science. She and Joshua Nall are also the co-authors of "Three-Dimensional Models: Uses and Representations", in Bernard V. Lightman (ed.), *A Companion to the History of Science* (Oxford, 2016).

Colour Plates

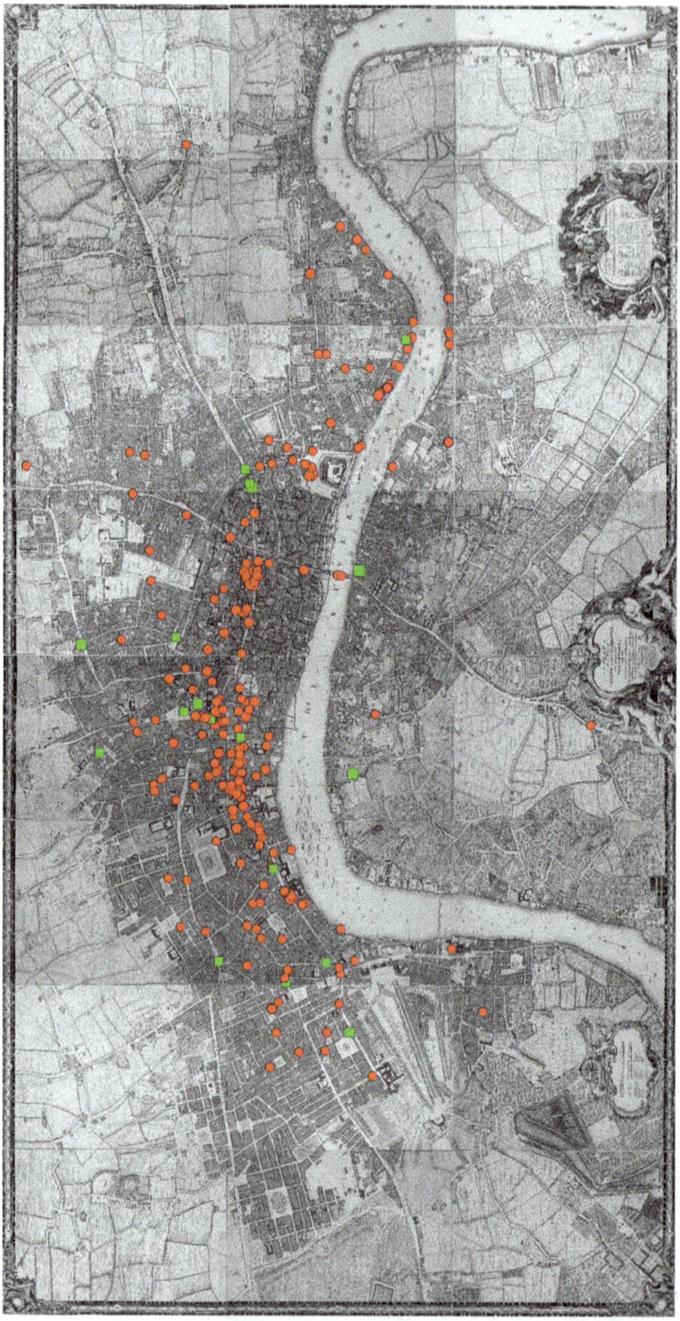

PLATE 1 *Digital map of shop locations known to have been involved in the London instrument trade during the period 1700–1750.*
UNDERLYING JOHN ROCQUE MAP OF 1746; COURTESY OF MOTCO ENTERPRISES LTD.

PLATE 2 *Portable telescope in black shagreen with ivory fittings and four drawtubes, made and signed in the early eighteenth century by George Willdey.*
GREENWICH, NATIONAL MARITIME MUSEUM, NO. NAV1522.

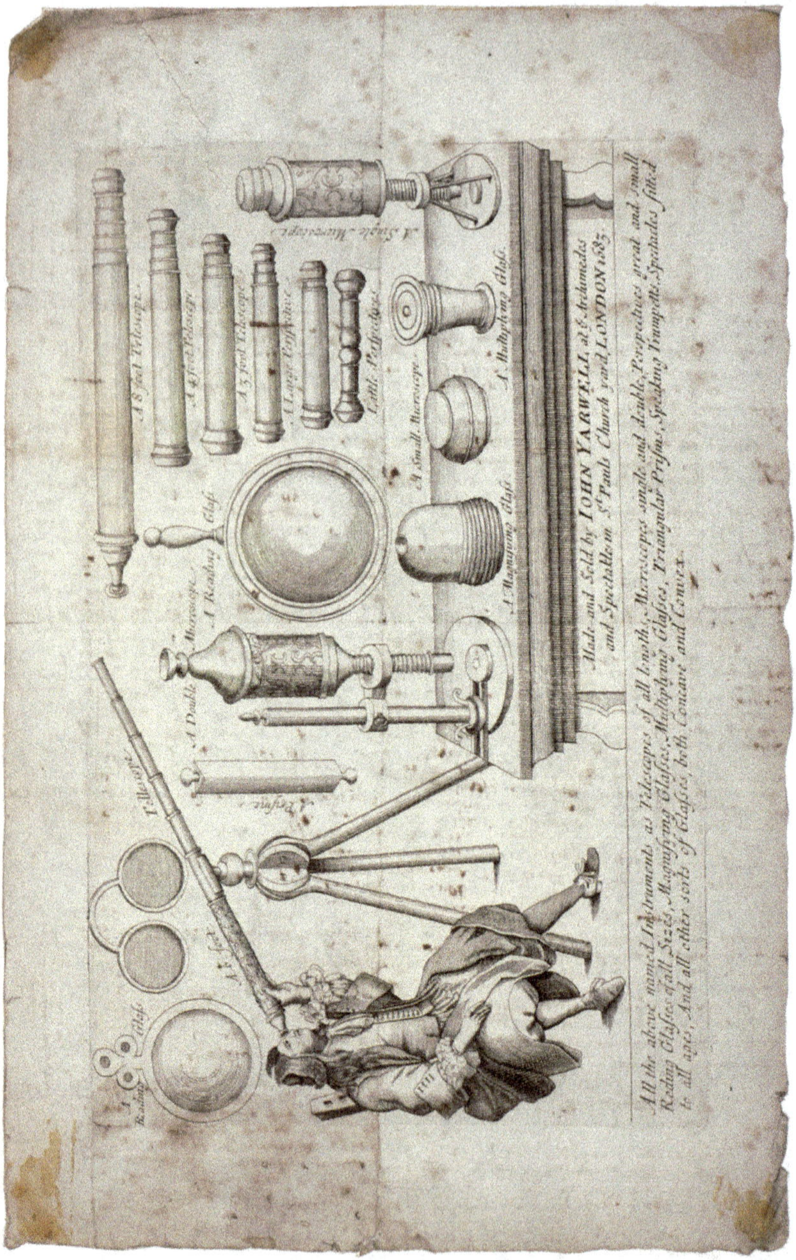

PLATE 3 *Trade card of John Yarwell, 1683. The printed sheet has manuscript notes by Isaac Newton on its reverse, and was purchased by Robert Whipple following a sale of Newton's papers in 1936.*
IMAGE © WHIPPLE MUSEUM, WH.3806.

PLATE 4 *Bohnenberger gyroscope, retailed by W. & S. Jones, London, c. 1835,* NMS *T.1990.26.*

PLATE 5 *Cary-Gould microscope, retailed by Alexander McKenzie, London, c.1820,* NMS *T.2008.170.*

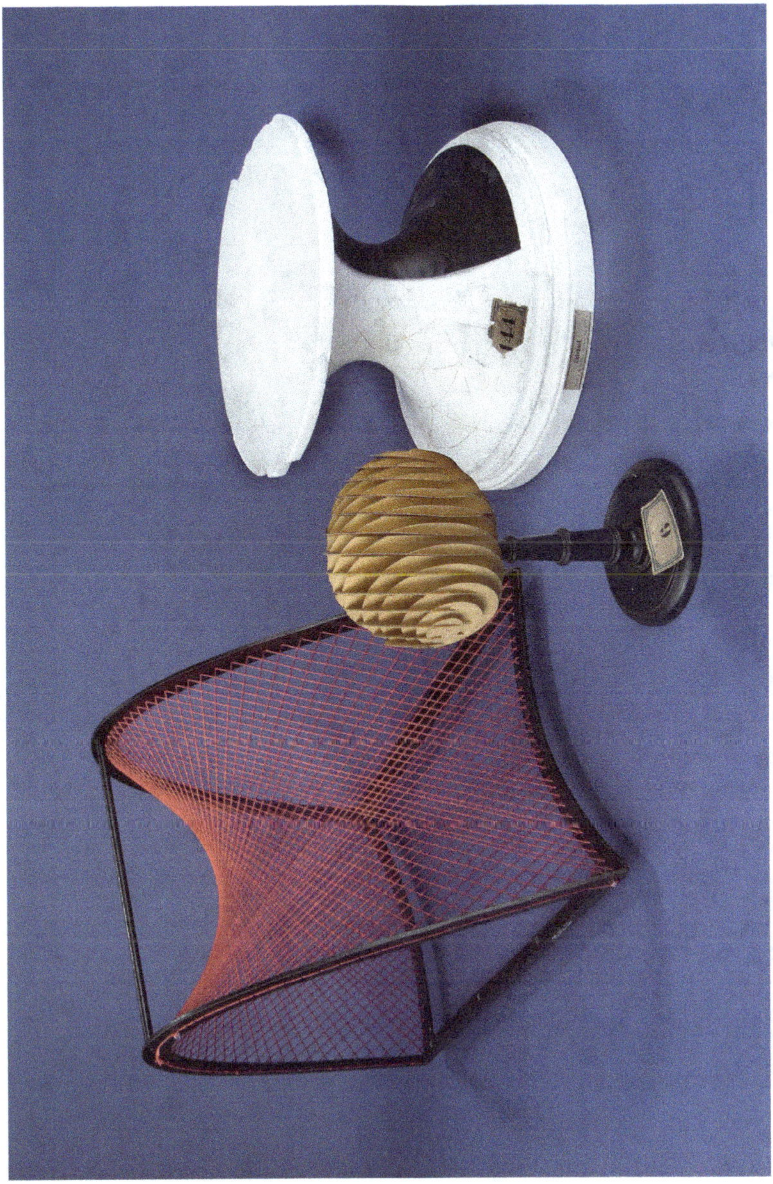

PLATE 6 *A few of the mathematical models by Ludwig Brill exhibited at and purchased from the World's Columbian Exposition in Chicago in 1893.*
SMITHSONIAN IMAGE 2006–2047.

PLATE 7 *Prototype Ritty cash register, exhibited by* NCR *at the Louisiana Purchase Centennial Exposition in St. Louis in 1904.*
SMITHSONIAN IMAGE 89-20453.

PLATE 8 *The depiction of a Native American and an Ottoman Turk on an ivory diptych by Thomas Ducher, Nuremberg, 1620–1640, tells us of its owner's fascinction with exotic peoples far from home.*

COLLECTION OF HISTORICAL SCIENTIFIC INSTRUMENTS, HARVARD UNIVERSITY, 7579.

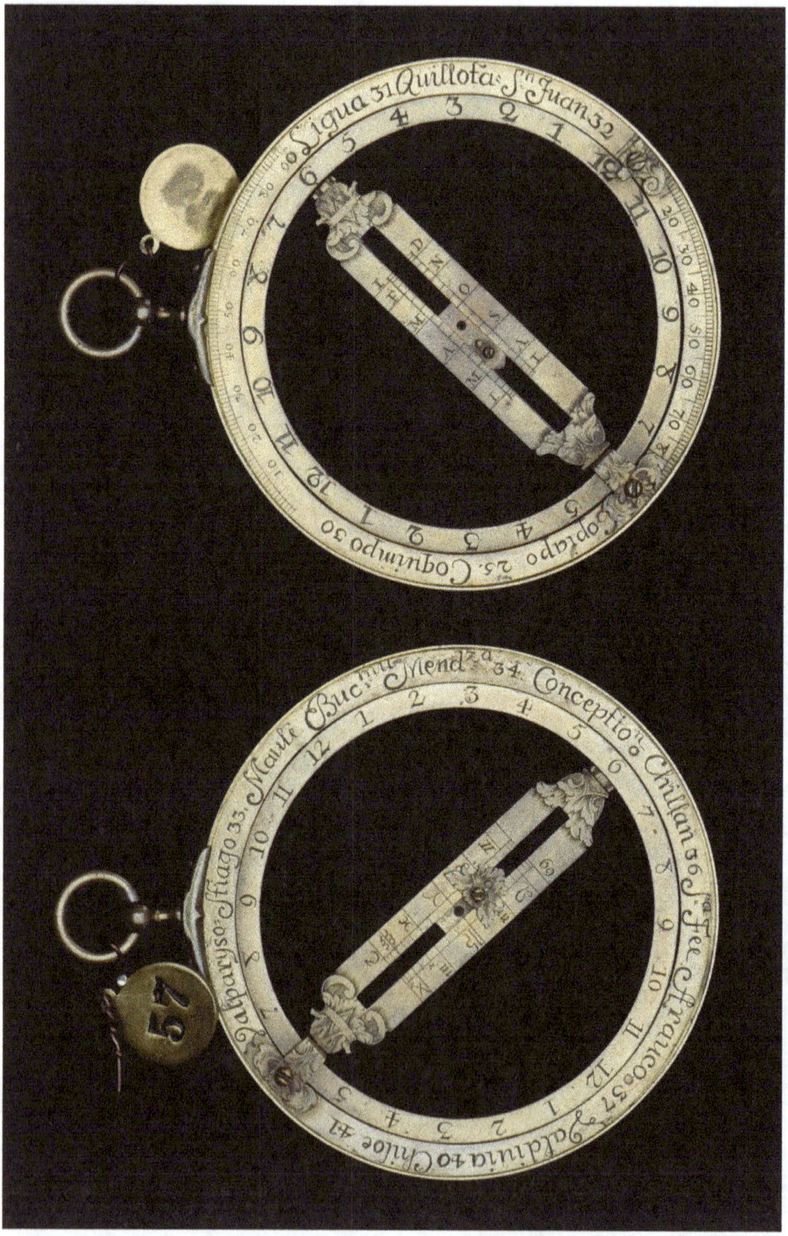

PLATE 9 *Universal ring dial made in Spain listing places in colonial Chile, ca. 1725–1775.*
COLLECTION OF HISTORICAL SCIENTIFIC INSTRUMENTS, HARVARD UNIVERSITY, 7904.

PLATE 10 *A sundial for New France by Pierre le Maire II, Paris, 1750–1760, with original case.*
STEWART MUSEUM, MONTREAL, CANADA, INV. 1988.51.

PLATE 11 *Inclining brass sundial for use along the Mississippi River down to the Gulf of Mexico, by Pierre le Maire II, Paris, 1730–1760.*
COURTESY OF ADLER PLANETARIUM, CHICAGO, ILLINOIS, W-57.

PLATE 12 *Alexander de Batz*, Native Americans of several nations in New Orleans, *1735, watercolor drawing. Gift of the Estate of Belle J. Bushnell.*

COURTESY THE PEABODY MUSEUM OF ARCHAEOLOGY AND ETHNOLOGY, HARVARD UNIVERSITY, CAMBRIDGE, PM NO. 41-72-10/20: DIGITAL FILE NO. 60741527.

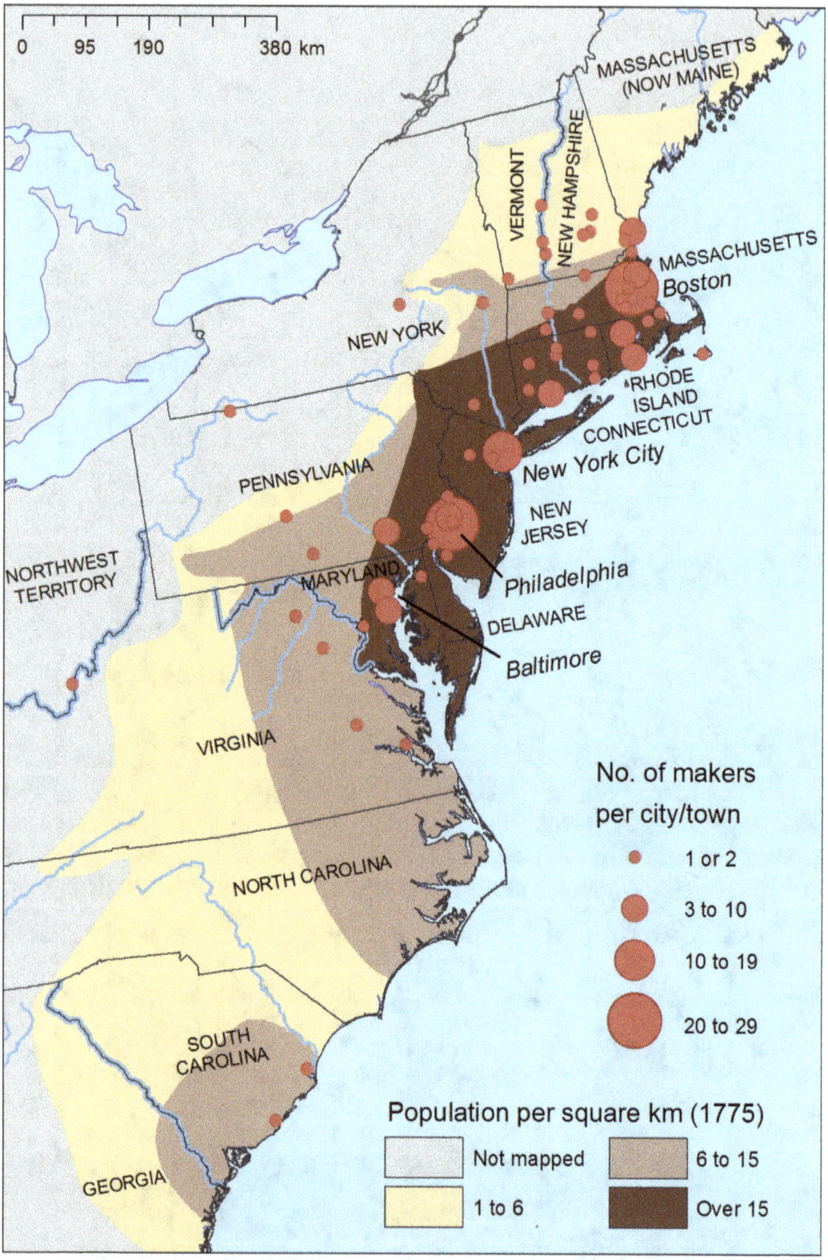

PLATE 13 *Location of makers in eighteenth-century North America, as listed in* Appendix.

POPULATION DENSITY DATA COURTESY US MILITARY ACADEMY, DEPARTMENT OF HISTORY; HISTORICAL BOUNDARIES COURTESY OF THE NEWBERRY LIBRARY; MAP ASSEMBLED BY JONATHAN W. CHIPMAN, DARTMOUTH COLLEGE DEPARTMENT OF GEOGRAPHY.

CHAPTER 1

Symbiosis and Style: The Production, Sale and Purchase of Instruments in the Luxury Markets of Eighteenth-century London

Alexi Baker

Introduction

In 1735, a popular play called *The Toy-Shop* premiered in London. Its author, Robert Dodsley, had previously worked as a footman and was thus familiar with the capital's stylish set. He used 'toys'—then fashionable trinkets for adults—including a telescope, spectacles, and scales as moralising metaphors. For example, when a young gentleman asks Dodsley's toyman about the precision of scales for making "statical Experiments", the shopkeeper convinces his listener that moral knowledge is of far greater importance than the scientific:

> *Toyman*: I have try'd them myself in some uncommon Subjects, and have prov'd their Goodness. I have taken a large Handful of Great Men's Promises, and put into one End; and lo! the Breath of a Fly in the other has kicked up the Beam. [...] I have found by Experience [...] that the Pride and Vanity of any Man are in exact Proportion to his Ignorance; that a Grain of Good-nature will preponderate against an Ounce of Wit; a Heart full of Virtue, against a Head full of Learning [...].
>
> *Gentleman*: Well, this is a Branch of Staticks, which I must own, I had but little Thought of entering into. However I begin to be persuaded, that to know the true Specifick Gravity of this Kind of Subjects, is of infinitely more importance than that of any other Bodies in the Universe.[1]

A modern audience might be somewhat surprised to find instruments included among fashionable 'toys' such as feather fans and agate snuffboxes, but they were constant companions to all manner of retail goods in early modern London. Globes, microscopes, pocket-sized mathematical and drawing tools, the occasional magic lantern or camera obscura, and especially vision

1 Robert Dodsley, *The Toy-Shop*, London, 1735, pp. 17–18.

 | DOI 10.1163/9789004324930_002

aids were widely sold alongside other stylish and luxurious wares. Spectacles were almost as common across all types of retail shops as were last-minute purchases like playing cards and lottery tickets.

Small fashionable instruments—and diversified luxury shops as points of sale for instruments—have received comparatively little attention in the history of science and technology. This probably stems from factors including the unrepresentative makeup of most surviving collections, and a certain prejudice against actors and technologies which might not today be deemed scientific and against shopkeepers assumed not to have made their own stock. However, small fashionable as well as workaday instruments accounted for a large proportion of the production and sales in early modern London (as in many other cities and towns). Furthermore, there were myriad socioeconomic interconnections between the different types of instrument makers and sellers—whether diversified and fashionable, or specialized and 'scientific'. The fruitful trade in instruments in the early modern capital can therefore not be fully understood without reintegrating fashionable and diversified commerce into the narrative.

Instruments in an Early Modern Metropolis

Instrument making first arrived in England from the Low Countries in the middle of the 1500s. As the number of local practitioners grew, its main focus became London—itself experiencing great geographic, demographic, and economic growth over the course of the early modern period. From about the late seventeenth to the mid-nineteenth centuries, London hosted more instrument makers and sellers than did any other location in the world, and its makers became the most esteemed in Europe alongside a small number of French artisans. This was at least in part because of comparatively light regulation by the government and livery companies and because of the existence of a well-developed consumer culture with diverse sources of demand.[2]

The products of this trade were not known as 'scientific', since the term 'science' did not take on its modern usage until the later 1800s. Enlightenment figures instead pursued subjects such as 'natural philosophy' and 'experimental

2 Gloria Clifton, *Directory of British Scientific Instrument Makers 1550–1851*, London, 1995, pp. xii–xiii; Maurice Daumas, *Scientific Instruments of the 17th and 18th Centuries and Their Makers*, London, 1989, pp. 246 and 258; Neil McKendrick, John Brewer and John H. Plumb, *The Birth of a Consumer Society: the Commercialization of Eighteenth-century England*, London, 1982, pp. 1–33.

philosophy', utilitarian projects such as the improvement of municipal water supplies, and a variety of topics from alchemy to religious chronology which would be considered beyond the purview of modern science. Eighteenth-century instruments were usually labeled and perceived as optical, mathematical, or philosophical. Most mathematical instruments had a graduated scale for performing calculations or for measuring angles or distance. Optical instruments involved lenses and glass or metal mirrors. Philosophical instruments were for the demonstration or investigation of natural phenomena such as magnetism, electricity, and the attributes of air.

Like many other early modern wares, these instruments were produced in ever-greater numbers and varieties during this period. This was in large part due to the expansion of consumerism which had begun by at least the seventeenth century, and because instruments could be geared towards so many uses and symbolic roles. They could be everything from precision technologies to everyday tools, to toys for adults and children (*Fig. 1.1*). For example, London instrument makers and retailers could sell globes, drawing instruments, and geometric models to educators; spectacles, planetaria, and electrical whimsies to the affluent and fashionable; telescopes, microscopes, and air pumps to researchers and observatories; and rules, compasses, and quadrants to the navy, military, trading companies, architects, navigators, and surveyors. More affordable versions of many of these wares could be crafted from paper, cardboard, and basic leather and then offered readymade. Inexpensive spectacles were even hawked in the street, as later described by Henry Mayhew, as were event-specific wares like eclipse glasses.[3] More expensive instruments could be offered either readymade or by commission in materials including valuable metals, shagreen, and inlays, or in more complex designs.

The existence of such a variety of uses and customers for instruments meant that they could symbolise many different things to different people as well, based upon attributes such as their uses and broader contexts of use but also their cost and style. Instruments communicated diverse meanings in physical form, whether viewed in use or static as when displayed in a study, and by their mention in texts from advertisements to plays. Instruments also communicated different meanings through their frequent appearance in visual contexts such as paintings and prints, shop signs, and even carved and moulded room décor. All levels of society would have been exposed to instruments in at least some of these visual contexts, since a number were public.

Whether in physical, textual or visual form, certain types of instrument could for example gesture towards the ideals and research activities of Newtonianism

3 Henry Mayhew, *London Labour and the London Poor*, London, 1861, v. 1, p. 444.

FIGURE 1.1 *Astronomy and use of the telescope were improving pastimes for the entire family.*
BENJAMIN MARTIN, *THE YOUNG GENTLEMAN AND LADY'S PHILOSOPHY*, LONDON, 1759: MUSEUM OF THE HISTORY OF SCIENCE AT OXFORD.

and the Enlightenment. However, expensive and stylish instruments could give the impression of genteel learning, fashion, and wealth as well. Pricy pocket-sized sets of mathematical instruments, or the use of a telescope disguised as a walking cane, could reveal as much about a person's grasp of fine taste and elegant movement as did the way in which he took snuff. Instruments could also represent cutting-edge professionalism for mathematical practitioners. For example, John Rocque advertised that the surveys for his well-known *Plan of the Cities of London and Westminster and Borough of Southwark* of 1746 were superior to past projects in part because he had employed a theodolite from the respected maker Jonathan Sisson.[4] Finally, instruments were frequently employed as common visual and textual metaphors based upon their uses. Optical instruments such as telescopes and microscopes were used to suggest qualities including insight and wisdom. Globes often gestured towards worldly power and knowledge, as in the '*memento mori*' paintings in which they were often used to represent fleeting human interests and accomplishments (in the face of the inevitable passage and decay of time as was represented by objects such as hourglasses and skulls).

The trade in these instruments in eighteenth-century London consisted of hundreds of shop-owning craftsmen and retailers (*Pl. 1*), who were enmeshed in the different socioeconomic communities and networks of the metropolis. There was as yet minimal industrialization of production, although improvements were made in materials and designs and in the precision dividing of scales during this period. Domestic workshops and/or retail shops remained the most common scenario, with only a small number of businesses employing large numbers of workmen onsite, as did that of Jesse Ramsden and that of the Troughtons—whose extensive workspaces are mentioned in the Old Bailey cases regarding a burglary at their premises in 1802.[5] Instrument makers usually relied upon the aid of a mixture of apprentices, journeymen, employees, and family members. This wider circle of participants may have numbered at least in the low thousands, especially if one factors in occasional workers and subcontractors.

It was common for instrument makers to buy or to barter for at least some components of their wares as well, as was true in many early modern crafts. For example, opticians often ground their own lenses and then subcontracted

4 Ralph Hyde, *The A to Z of Georgian London*, London, 1981, pp. v–vii.

5 Anita McConnell, *Jesse Ramsden (1735–1800) London's Leading Scientific Instrument Maker*, Aldershot, 2007, pp. 53–75; Trial of William Bean (t18020217-70), February 1802, and Trial of John Bazley (t18020428-60), April 1802: *Old Bailey Proceedings Online* (www.oldbaileyonline.org, version 7.2, accessed 17 February 2016).

FIGURE 1.2 *Nathaniel Hill (died 1768), and likely before him John Coggs, advertised all three classes of instrument with this trade card.*
OXFORD, MUSEUM OF THE HISTORY OF SCIENCE, JR MILLBURN BEQUEST.

for frames, tubes, and fitted cases since their production required different skills and materials. It also became increasingly common as the eighteenth century progressed for shops to offer all three classes of instrument—optical, mathematical, and philosophical (*Fig. 1.2*)—and to expand to include other crafts and trades as well, often through subcontracting and bartering. These included metal (e.g., clockmaking or cutlery), luxury (e.g., toys), mathematical (e.g., surveying or education), maritime (e.g., shipchandlery or navigation), and print trades. It was a rare maker or retailer who could flourish as a narrow specialist, as did the highly successful telescope mirror maker James Short. As a result, instruments became common stock across the diverse shops of the metropolis.

Technologies as 'Toys'

The retail sale and export of instruments as stylish luxuries gathered momentum with the general growth of consumer culture in early modern London and other large British cities. By the 1700s, browsing the contents of shops had become a popular pastime for genteel individuals. This was encouraged in part by developments such as the glazing of shop fronts in order to display wares to passerby to the best advantage, and the opening of large shopping venues such as the second version of the Royal Exchange built in 1669. Shoppers could stroll a prime retail street and peruse the contents of the premises of instrument makers, print sellers, toymen, jewellers, and mercers. 'Toys' were expensive and highly popular trinkets intended for adults such as mother of pearl game counters, ingenious snuffboxes, small accoutrements, and household décor. The moral philosopher Adam Smith quipped about the lovers of these goods that, "All their pockets are stuffed with little conveniences. They contrive new pockets, unknown in the clothes of other people, in order to carry a greater number".[6] Such merchandise was displayed to the best advantage through gleaming windows, and inside shops in glass cases and on polished wooden counters. A fine display of stock also implied that a retailer could obtain more high-quality wares for his or her customers.[7] When the optician and toyman George Willdey renovated his premises in Ludgate Street in 1712, he had a glazier mend the counter lantern and refresh the panes of glass in his window and cases in order to optimize the exterior and interior display of wares. A painter silvered and lacquered the shop fittings and furniture, and lacquered a number of instruments while he was there.[8] There was also "ye spectacle glass Case on ye stall", which may refer to a glazed case of spectacles set up on a table outside of the shop in order to attract customers, with the table hinging up to shutter the building at night.[9]

Diversification increased in retail shops of all kinds as part of the early modern surge in consumerism, and later from increased industrialization and the beginnings of mass production. All manner of textual and pictorial advertising and commercial and consumer records show that small instruments were a frequent element of this trend, with vision aids and pocket sets

6 Adam Smith, *The Theory of Moral Sentiments*, London, 1767, p. 266.

7 Jon Stobart, Andrew Hann, Victoria Morgan, *Spaces of Consumption: Leisure and Shopping in the English Town, c. 1680–1830*, Abingdon, 2007, pp. 116, 126 and 130; Kathryn Morrison, *English Shops and Shopping*, New Haven, 2003, p. 35.

8 "Fenhoulet v. Willdey", 1744: London, The National Archives, C 104/21, Book R.

9 Ibidem; Morrison, op. cit. (n. 7), p. 24.

of mathematical instruments especially popping up among otherwise unrelated wares. Spectacles were immensely popular, being both everyday utilitarian tools and fashion statements, which could be manufactured to affordable or luxurious specifications. Moses Jacobs was both a spectacle maker and a dealer in hardware "next door but one to the Synagogue in Duke Place in the parish of St. James".[10] The cutler John Brailsford advertised spectacles, rules, and drawing instruments near Leicester Fields at mid-century.[11] Trade cards for the razor maker Henry Patten of Holborn from the same time, depict two styles of spectacles among a wide variety of cutlery and fishing tackle.[12] As a final example, multilingual trade cards for Walter Robotham near the Royal Exchange mention spectacles and telescopes ("Prospective Glasses") alongside "All sorts of Curiosities in Gold, Silver, Amber, Tortoise-Shell, Ivory, Steel, & all other Metals".[13]

Sometimes early lens grinders, both commercial and gentleman amateurs, warned that vision aids should not be purchased from these diversified retailers. The well-known optician John Marshall—perhaps in part to take a swipe at his former workman George Willdey—advertised that "Abundance of Peoples Sight is spoiled by using false Spectacles, sold by Toymen and other ignorant Persons".[14] This refrain was mimicked in the satirical pamphlet, *A true system of religion, according to the best astronomical observations* of 1722. In it an insane young Cambridge scholar advised readers to use the "Telescope of Reason" to view (somewhat erotic) religious allegorical truths by grinding the lenses for it themselves, "till you have fitted them to your own Sight, and by no means to trust to the idle Toys made by common Mechanics".[15] Dodsley's *Toy-Shop* similarly derided the emphasis on fashion rather than efficacy when it came to spectacles.[16] However, in reality, a number of London toymen and many wholesale metropolitan suppliers of vision aids had actually trained in the optical instrument trade.

Many London toyshops sold instruments, and a number of instrument makers expanded their retail businesses to include toys. Some examples

10 "Policy of Moses Jacobs" no. 95963, 19 August 1743: London, Museum of London, Sun Insurance 11936, 66, p. 338.

11 Clifton, op. cit. (n. 2), p. 37; Ambrose Heal, *London Tradesmen's Cards of the XVIII Century*, New York, 1968, p. 73 and pl. xxiv.

12 "Henry Patten trade card": London, Victoria & Albert Museum, no. E.571-1976.

13 "Walter Robotham trade card": London, Victoria & Albert Museum, no. E.2376-1987.

14 *Daily Post*: 17 February 1721, 24 March 1721, 23 January 1722.

15 Phil. Alethson, *A true system of religion, according to the best astronomical observations*, London, 1722, p. 7.

16 Dodsley, op. cit. (n. 1), pp. 32–33.

include George Willdey and his successors, George and Mary Sterrop, Samuel Saunders, Thomas Wright, Benjamin Gray, Christopher Pinchbeck, and the retailers Marrianna Viet and Thomas Mitchell. Most lined the important retail corridor of Fleet Street, Ludgate Street, and St. Paul's Churchyard. Willdey, the Sterrops and later Saunders all clustered at St. Paul's alongside the longstanding print traders and other luxury retailers. Viet and Mitchell, on the other hand, operated near the Royal Exchange, which was a key retail centre for both the instrument trade and for the capital at large.

A number of these individuals were aided in their forays into different metalworking and luxury trades by relationships with the French immigrant community. Most but not all of these immigrants were Huguenots, or Calvinists who had been victimized at home by the Catholic authorities after the revocation of the Edict of Nantes in 1685. New arrivals from France could draw upon extensive networks of family, friends, and acquaintances who were already in England and who sometimes guided them into the instrument trade from similar crafts.[17] These included printing, engraving, metalworking, clock making, and overseas trading. The actual instrument trade in France was comparatively weak due at least in part to more rigorous regulation, a less vital and flexible consumer economy, and a significantly different organization of society. Native Britons could equally benefit from these networks through marriage into the French community.

For example, the mathematical instrument maker Thomas Wright married the French Protestant Suzanne de la Balle and took advantage of her connections for the toy side of his business.[18] Wright, Willdey, and Mitchell also married into the French community—unless Mitchell was an Anglicized name—while Marrianna Viet was a Huguenot herself. Trade cards reveal that Viet and Mitchell sold spectacles as well as toys and jewellery at the Dial & Kings Arms on Cornhill near the Royal Exchange.[19] Marrianna was the widow of Claude Viet, a "deceased Watchmaker to her late Majesty Queen Caroline,"

17 Irene Scouloudi (ed.), *Huguenots in Britain and their French Background, 1550–1800*, London, 1987, p. 28; Robin Gwynn, *Huguenot Heritage: The history and contribution of the Huguenots in Britain*, London, 1985, pp. 1, 3–4, 21–23, 36, 60, 71 and 86–87.

18 It is possible that Suzanne was also the Susanna Ball, daughter of Lewis, apprenticed to optician George Willdey in 1715. "Will of Susanna Wright", 16 February 1769: London, The National Archives, PROB 11/946; "Will of Lewis De La Balle", 7 November 1758: London, The National Archives, PROB 11/841; "Spectaclemakers' Archives": London, Guildhall Library, MS 5213-1; William Minet, "The de la Balle family of Guînes", *Proceedings of the Huguenot Society of London* 10 (1912), pp. 98–129.

19 Heal, op. cit. (n. 11), pl. 9.

and mother-in-law to Mitchell.[20] Their business continued in operation until 1758, beyond the deaths of its founders. Either Marrianna's daughter or niece was the first woman admitted to the Clockmakers' Company as an apprentice.[21]

The French represented the single largest and most influential immigrant group in the instrument trade of early eighteenth-century London. Seven or more percent of trade members may have had recent French origins, although this is difficult to quantify given the Anglicization of names and the incomplete historical record. The majority were involved with optical instruments or spectacles, and thus entered the Spectaclemakers' Company, while a smaller number were mathematical instrument makers and Clockmakers. These trade members seem to have largely come from similar family crafts, although there were exceptions including the well-known optician Peter Dollond, who originally worked as a silk weaver.

Among these individuals, George Willdey has left behind the greatest amount of surviving archival records detailing his business. His marriage, and that of his daughter, into the French Huguenot community allowed him to expand his business from solely making and selling optical instruments to offering diverse luxury wares. George Willdey and his family operated a highly successful optical and toy shop near St. Paul's Churchyard for forty-two years. Their extant shop accounts are the only ones known to have survived from the first 200 to 250 years of the British instrument trade.[22] As a result, these records not only dramatically illuminate the intersections between instruments and fashion, but also the London instrument trade as a whole.

"The Most Noted Toyman in Europe"

George Willdey was remarkable among instrument makers.[23] He was by far the longest-serving Master of the Spectaclemakers' Company during the early modern period, holding the position for eleven years after having already

20 "Will of Thomas Mitchell", 2 January 1752: London, The National Archives, PROB 11/792; "Will of Claude Viet", 4 March 1734: London, The National Archives, PROB 11/664; "Will of Mary Anne Viet", 24 April 1744: London, The National Archives, PROB 11/733.

21 Philip J. Shears, "Huguenot connections with the clockmaking trade in England", *Proceedings of the Huguenot Society of London* 20 (1959–1960), pp. 158–176: 167–168.

22 Alexi Baker, *"This Ingenious Business": The socio-economics of the scientific instrument trade in London, 1700–1750*, PhD dissertation, University of Oxford, 2010, pp. 254–321.

23 Alexi Baker, "Willdey, George (bap. 1676, d. 1737)", *Oxford Dictionary of National Biography*, Oxford, 2015; www.oxforddnb.com/view/article/50906 (accessed 28 July 2016).

served on the Court for twelve years. The optician became well-known across Britain and much of Europe for his fashionable retail and wholesale business, which he oversaw from 1706 until his death in 1737 and which then continued under his wife and son until 1748. In part, Willdey achieved this renown through far greater use of newspaper advertising than any other instrument maker. The shop was name-checked in publications and received many affluent, titled, and royal visitors from home and abroad. The toyman also conducted large-scale wholesale transactions with a wide range of tradesmen across Europe including those in Portugal, Spain, Italy, Germany, and the Netherlands.

However, Willdey's surviving accounts reveal that he also engaged in socio-economic practices and relationships which were common to the preindustrial trade as a whole. These included frequent subcontracting and bartering both within and without the trade, extensive reciprocal ties to members of other crafts and retail specialties, and mutually beneficial memberships in metropolitan communities including those of the French Huguenots and the Spectaclemakers' Company. Willdey was also one of many trade members and diversified retailers across Britain and Europe who capitalised on the widespread popularity of instruments, and particularly of small stylish vision aids. He moved thousands if not tens of thousands of instruments and instrument components, including by supplying other members of the London trade and by extensively using his optical wares to barter for other types of stock.

Willdey, born the son of a 'gentleman' of Staffordshire, made his way to London after his father's death. On 25 February 1695, he was bound in the Spectaclemakers' Company to John Yarwell, a prominent and commercially successful optician who had recently served as its Master and was located in Ludgate Street near St. Paul's Churchyard.[24] This neighbourhood, where the new apprentice would end up spending his entire working life, was a desirable mix of the longstanding print trade and of luxury retailing. Willdey may have been a slightly late-blooming seventeen or eighteen at this time.[25] He gained his freedom from the livery company on 26 March 1702, and worked as a journeyman for Yarwell and then for the well-known lens maker John Marshall, also located nearby.[26] Within four years, the young man was able to meet the start-up costs of opening his own business by entering into partnership with Timothy Brandreth, who had served an apprenticeship with Ralph Sterrop and then also worked for Marshall as a journeyman.

24 "Spectaclemakers' Archives", London, Guildhall Library, MS 5213-1.

25 Based on his stated age when writing his will, "Will of George Willdey", 16 November 1737: London, The National Archives, PROB 11/686.

26 "Spectaclemakers' Archives", London, Guildhall Library, MS 6031-1.

Willdey and Brandreth operated under the sign of the Archimedes and Globe near the Dog Tavern in the middle of Ludgate Street, and used newspaper advertisements and trade cards to offer the public a wide range of spectacles and optical instruments. These included a two-foot telescope which they said could tell someone "the Hour of the Day 3 if not 4 Miles by such a Dial as St. James's or Bow"; a "Concave Metal Speculum that will unite the Sun Beams so vigorously that it will melt Steel and viterfy [sic] the hardest Substance"; and "several sorts of Microscopes, some of which magnifies the object above a Million of times".[27] Following the example of other trade members, they also periodically claimed to be writing "a small Treatise, with the Advice and Assistance of the Learned, that gives the Reasons why they do so; and other Curiosities, which will speedily be finished, and given (Gratis) to our Customers".[28]

It is highly likely that the partners ground their own lenses but then subcontracted for tubes and frames and cases, as Willdey's later shop accounts imply was the case. Yarwell and Sterrop, who entered into a brief but vituperative newspaper advertising war with their former apprentices in 1707, tried to use this as an excuse to accuse them of advertising microscopes that had actually been made by the well-known mathematical instrument maker Edmund Culpeper. It is likely that there was a reciprocal relationship—not unusual at the time—by which the microscopes sold at both businesses had optics made by the partners and fittings by Culpeper. The accounts show that Willdey went on to have similar relationships of subcontracting and barter with many other instrument makers. Within a few years, he and Brandreth were openly advertising their sale of mathematical instruments "all made to the greatest Perfection by *Edmund Culpeper*".[29]

In November 1709, George married Judith Sene or De Senne, a 30 year-old spinster and French Huguenot immigrant of St Anne's Blackfriars.[30] Over the years Judith's extended family and friends would provide Willdey with many suppliers of diverse luxuries for resale at his shop and a number of his fifteen apprentices, as well as agents and customers and some loans. These resources greatly facilitated the retailer's expansion into the toy, jewellery, and

27 *Daily Courant*, 29 April 1707. The partners' sign closely resembled both the Archimedes and Three Spectacles employed by their former masters, and the Archimedes and Spectacles used by their former employer John Marshall.

28 *English Post with News Foreign and Domestick*, 19 March 1706.

29 D. J. Bryden, "Evidence from Advertising for Mathematical Instrument Making in London 1556–1714", *Annals of Science* 49 (1992), pp. 301–336: 326.

30 Brian Beet, "Willdey's Toyshop", *Proceedings of the Huguenot Society of Great Britain and Ireland* 28 (2003), pp. 92–94: 92.

other luxury metalworking trades. The diversified retailing and Huguenot ties further contributed to an unusually high degree of official and front-of-house involvement in the business by women. These included Willdey's wife Judith, apprentice turned long-time shop manager Susannah Passavant, and seven other female apprentices.[31]

By 1709, Willdey and Brandreth began to expand their stock to include maps and soon globes made by John Senex and Charles Price and their associates.[32] By the next year, the Ludgate Street partners were advertising globes in diameters of three, twelve, and sixteen inches, and they had erected a large pair of the spheres on posts in front of their shop as additional advertising.[33] George would continue to sell maps and prints and occasionally globes throughout his career, and he both commissioned new maps and atlases and purchased and reused the copper plates for popular maps of past decades.[34] This included an atlas of the world, begun in 1714 and finished in 1718, which was decorated with dense pictorial advertisements for instruments and toys (*Fig. 1.3*).[35] By at least 1710, he and Brandreth had obviously begun to expand into the other metal trades as well, since the newspaper notice of a break-in at their shop mentions metal utensils and kitchen implements, buckles and seals, and a snuff box.[36]

In 1711, Brandreth departed for the equally desirable neighbourhood of Cornhill near the Royal Exchange, where he established a shop selling the same type of wares under the same sign of Archimedes and Globe.[37] There is no extant evidence for his having achieved as much commercial success, perhaps in part because he died sometime between 1715 and 1720.[38] At this

31 Many women were involved in the early modern instrument trade as facilitators, and some were shop owners in their own right—sometimes for years or even decades—but they were rarely official apprentices.

32 *Post Man and the Historical Account*, 8 September 1709; "Fenhoulet v. Willdey", op. cit. (n. 8).

33 *Lucubrations of Isaac Bickerstaff*, 25 February 1710.

34 Willdey obtained the copper plates for Christopher Saxton's atlas of England and Wales of 1574–1579 after Anne Lea died and her partner Richard Glynne retired in 1730, and published 33 of those maps as the atlas *The Shires of England and Wales*. Edward Lynam, *The Mapmaker's Art*, London, 1953, p. 89; Tony Campbell, "Chronicle for 1980", *Imago Mundi* 33 (1981), pp. 108–114: 111.

35 George Willdey, "Toy atlas", 1714–1718: London, National Maritime Museum, no. 3793-2001.

36 *Daily Courant*, 28 July 1710.

37 Ibid., 7 July 1711.

38 The Spectaclemakers' Archives last mentioned Brandreth in 1715, and his wife was last mentioned by name trading at the same location the following year. She may have later remarried and been the Mrs. Guy who operated at the same location from 1720 to 1727. "Spectaclemakers' Archives", London, Guildhall Library, MS 5213–1; *Post Man and the Historical Account*, 18 September 1716. *Weekly Journal or Saturday's Post*, 31 December

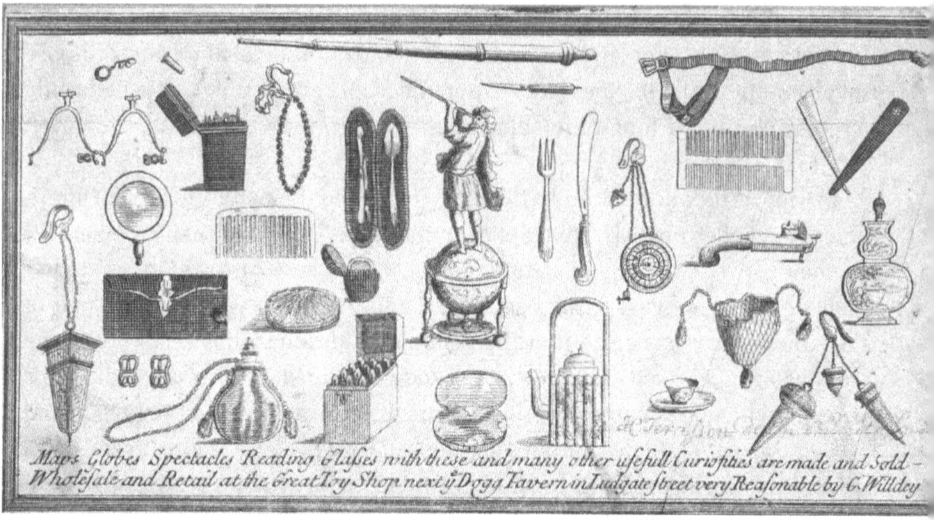

FIGURE 1.3 *An early map insert advertising George Willdey's shop, also reproduced elsewhere as a freestanding trade card.*
GEORGE WILLDEY, "TOY ATLAS", 1714–1718: GREENWICH, NATIONAL MARITIME MUSEUM, NO. 3793-2001.

time, Willdey took out a fire insurance policy with Sun Insurance, then only in its second year of operation and still operating out of a rented room at Causey's Coffee House in nearby St. Paul's Churchyard. The policy described him as a spectacle maker and toy trader in Ludgate.[39] The optician remained in this shop until 1716, selling spectacles and instruments and a wide variety of other metal, printed, and luxury goods.

The first notice of Willdey's beginning the move to a second shop came in September 1715, when an advertisement mentioned that he was occupying "both the Print and Map Shop the Corner of Ludgatestreet next St. Pauls, and [...] the Toy shop next the Dog Tavern in the same Street".[40] By the following July, he was '"removed [...] to the Corner House a little higher next St. Paul's".[41]

1720; *Daily Post*, 11 May 1727. The Whipple Museum in Cambridge possesses a telescope by Brandreth, which is similar in appearance to the surviving Willdey telescopes.

39 Peter Dickson, *The Sun Insurance Office 1710–1960*, Oxford, 1960, pp. 31 and 83–84; "Policy of George Willdey" no. 1603, 25 December 1711: London, Museum of London, Sun Insurance 11936, v. 1, p. 163.

40 *Post Man and the Historical Account*, 8 September 1715.

41 Ibid., 7 July 1716.

The advertisements indicate the optician's gradual relocation to the shop previously owned by the map and print seller Christopher Browne, and before that by Browne's former master Robert Walton, which was on the very corner of Ludgate and opened out onto a grand view of St. Paul's.[42] At this second location, Willdey continued to greatly expand his retail and wholesale business until his death in 1737.

Throughout his career, Willdey issued hundreds of newspaper advertisements—an unprecedented number for the instrument trade, and rare even for luxury retailers of the time—as well as pictorial trade cards and map and print inserts. In 1729, he boasted in a newspaper that "no shop in Europe can shew so great a Choice as his contains".[43] Willdey often exhibited a keen sense of showmanship, which he had first displayed in public during the newsprint conflict with his former employers in 1707. Yarwell and Sterrop then called the salesman's "foul Language no better than Billingsgate Railing"—in other words, the strident harangue of the fish hawkers at Billingsgate Market. The younger man in turn accused his former master and employer of jealousy and deceit and challenged them to public comparisons of their respective wares, with ever-greater wagers place on the outcome.[44]

Willdey was never again so confrontational in public, but he did continue to exhibit an entrepreneurial flair. For example, thirteen years later, he advertised the erection of a giant burning glass on the roof of his house as a tourist attraction that referenced actual scientific experiments, which could reportedly handle everything from culinary boiling and baking to the melting of metals: "It and its surprising Effects are shew'd Gratis to any of my Customers, that lay out Five Shillings, or more with me; provided the Sun shines, and the Air be Clear. N.B. This far exceeds that show'd in the Privy Garden in White Hall, though each Person paid Half a Crown for the Sight of that".[45]

No matter what ventures Willdey embarked upon, and no matter how much he diversified his stock, optical instruments always remained central to his self-identity. They also continued to represent a significant chunk of the retail and wholesale goods that he moved each year, at least until the illness that preceded his death. The most complete of his surviving shop accounts are for 1712 and 1713, and then instruments accounted for at least 29 and 25 percent

42 Willdey may have been assisted in funding the move by the £100 bequest that he received in 1715 from his wife's brother-in-law, Peter Ferment. "Will of Peter Ferment", 3 March 1715: London, The National Archives, PROB 11/545.

43 *Daily Post*, 30 January 1729.

44 *Daily Courant*, 29 April 1707.

45 *Post Man and the Historical Account*, 22 October 1720.

respectively of the value of the stock moved—and much more than that in terms of the numbers of items sold or bartered. Optical instruments, and the tubes for making microscopes and telescopes, were not only some of Willdey's best-selling wares. They were also, even more so than his other types of stock, the currency with which he bartered with other craftsmen in order to be able to boast of carrying the greatest variety of goods in Europe.

A wide range of tradesmen, including other London instrument makers, recognized that their customers were also likely to purchase the optician's decorative instruments and added them to their stock, often in return for the goods which they made. These wholesale customers ran the gamut from makers of apparel and jewellery and cutlery and toys, to those of maps and books and maritime goods. Among the associates from within the London instrument trade were rule maker Mr Peack Jr., spectaclemaker William Saunders, mathematical instrument maker Samuel Saunders, spectacle and mathematical instrument maker William Arundell, optician and glass grinder Thomas Ribright the elder, possibly rule seller Joseph Silvester and instrument maker Benjamin Hickman, and possibly the optical instrument maker George Bass and spectacle maker Thomas Lincoln (who later valued Willdey's stock for probate). This is only a partial list, given the fragmented survival of the shop accounts.

In addition to these and numerous other London customers, there were many (largely wholesale) customers located in provincial towns including Windsor, Plymouth, and Colchester. Willdey served large wholesale customers overseas as well including some in Portugal and Spain and perhaps in what are today Italy, Germany, and the Netherlands. Attractive instruments were clearly popular in many European capitals, just as they were in Britain. The optician's incomplete surviving shop accounts, which provide spotty coverage from across the period of 1710 to 1725, reveal that he was selling and bartering thousands of instruments during the early eighteenth century. Since partial accounts alone provide a tally of almost 12,000 finished instruments and thousands of instrument components (*see Table 1.1*) for the most popular types, this may suggest that he sold and bartered tens of thousands of instruments across the entirety of his career.[46]

46 "Fenhoulet v. Willdey", op. cit. (n. 8).

TABLE 1.1 *Sale and barter estimates for the most popular instruments in George Willdey's partial surviving shop accounts from 1710–1725 ("Fenhoulet v. Willdey", 1744: London, The National Archives, C 104/21, Book R)*

	Tubes	Spectacles	Telescopes	Burning Glasses	Microscopes	Reading Glasses
1710	N/A	N/A	N/A	N/A	4	N/A
1711	14	54	14	N/A	8	13
1712	2790	1921+	734	513	118	169
1713	1127	927	288	87	47	57
1714	938	580	282	13	35	33
1715	312	495	503	194	221	13
1716	36	102	185	N/A	102	36
1717	90	114	274	48	18	15
1718	N/A	184	171	37	72	10
1719	N/A	156	150	N/A	72	7
1720	N/A	6	N/A	N/A	N/A	12
1721	N/A	12	N/A	1	N/A	2
1722	156	637	888	294	128	21
1723	264	102	268	N/A	36	74
1724	N/A	N/A	N/A	N/A	N/A	N/A
1725	N/A	18	1	N/A	N/A	7
Total	5727	5308+	3757	1187	861	469

For example, the records show Willdey selling and bartering almost 3800 telescopes, from a two-inch model intended for the short-sighted to one 20-foot focal length instrument. A number of the optician's large wholesale customers bought a representative array of these different sizes for their own shops. Each one could cost anywhere from shillings to a small number of pounds, depending upon size and materials. Willdey offered telescopes in a myriad of different combinations of materials including vellum, brass, silver, ivory, fishskin, tortoiseshell, and woods such as mahogany and walnut. The instruments could be dyed, painted, or constructed in colours including green, black, white, red, marbled, or straight metal or wood. They could also be decorated with

FIGURE 1.4 *Objective lens cap signed "G. Willdey Fecit" from early eighteenth-century telescope by George Willdey in black shagreen with silver fittings and three drawtubes.*
KASSEL, MUSEUMSLANDSCHAFT HESSEN KASSEL, NO. F321.

inlay or with additional turner's work and were sometimes mentioned as being lacquered or Japanned. Once a body was crafted in these materials and styles, it could also have varying numbers of coloured and gilt drawtubes installed.

So far the only known surviving instruments by Willdey are two portable telescopes bound in black shagreen, with green and gold leather drawtubes. That now at the National Maritime Museum in Greenwich is missing its lenses but has ivory rings and four drawtubes, one of which has "George Wildey f M.X" handwritten around its base (*Pl. 2*).[47] The other telescope in the Museumslandschaft Hessen Kassel in Germany still has its lenses, three drawtubes, and more expensive silver fittings including an objective lens cap that is signed "G. Willdey Fecit" (*Fig. 1.4*).[48] It is possible that it was bought for use at the college that briefly existed at the Castle. These instruments are also of

47 George Willdey, Telescope, *c.* 1710: Greenwich, National Maritime Museum, no. NAV1522.
48 George Willdey, Telescope: Kassel, Museumslandschaft Hessen Kassel, no. F321.

particular interest because the two-foot model of telescope was often central to Willdey's advertising and self-presentation, beginning with the public fight in 1707 over its quality as compared to the comparable model sold by Yarwell and Sterrop.

By the end of Willdey's life, the sale of copious optical instruments, of occasional mathematical and philosophical instruments, and of complementary luxuries had garnered the optician remarkable socioeconomic success and standing. His shop was well-known among the fashionable and affluent of Britain and Europe and received many titled and even royal visitors, as when the *Daily Post* announced in 1736 that the Prince and Princess of Wales had visited.[49] The Willdeys themselves were sometimes discussed in the society pages alongside their well-heeled customers. This social and commercial success was reflected in the value of the business and properties left in 1737. Despite the possible contraction of George's wholesale efforts in later years due to illness, his stock at his death was worth about £ 7,014. This included almost 1700 spectacles, 350 telescopes, and many burning glasses, microscopes, tubes, and other components. These continued to sell, when the shop reopened under his widow Judith Willdey and shop manager Susannah Passavant.[50] George also left behind two comfortable homes, fashionable trappings including a carriage, and an estate worth more than £ 9,000 altogether.[51] It was most common during this period for the estates of instrument makers to be worth tens or hundreds of pounds, although the wills of other trade members who were also toymen hint at wealth in the thousands of pounds though often without specifying monetary amounts. Of course, an instrument maker's estate at the time of death did not always reflect the state of his business in years past. The optician Oliver Combs once successfully operated shops near Leicester Fields and in the Strand but appears to have been in dire straits by the time he died, possibly aged in his eighties, only able to leave bequests of a shilling each to family and clearly unsure that even a future as servants could be secured for his youngest daughters.[52]

Among the tributes that the newspapers ran after George Willdey's passing, *The Old Whig or The Consistent Protestant* heralded the optician as "the

49 *Daily Post*, 18 September 1736.

50 "Probate inventory of George Willdey", 1 July 1738: London, The National Archives, PROB 3/37/71.

51 Ibid.; "Fenhoulet v. Willdey", op. cit. (n. 8); "Will of George Willdey", op. cit. (n. 25).

52 "Will of Oliver Coombes or Combs", 4 February 1752: London, The National Archives, PROB 11/792.

most noted Toyman in Europe".[53] The *London Evening Post* noted that he had "acquir'd a plentiful Fortune with a fair Character", and *The Daily Post* and *Daily Gazetteer* both reported that he was "said to have died very rich".[54] His reputation would linger even longer, although the shop itself only endured for eleven more years until the death of the family's only surviving son. As late as 1750, the poet Mary Jones would still be writing nostalgically of withstanding the "temptations thick and strong [...] [to] stop At Wildey's toys".[55]

Willdey's life and those of many of his colleagues argue for greater attention to be paid in the history of science and technology to diversified and luxury retailers such as toymen. They moved vast quantities of instruments across Britain and Europe, in some cases having also made the instruments themselves. They were furthermore often tied to other members of the instrument trade by reciprocal behind-the-scenes relationships of subcontracting and barter. The pervasive incorporation of instruments into diversified retailing reflected their widespread popularity and utility, and the frequent intersection of instrument making with other crafts and trades only strengthened the trade as a whole by giving its members greater socioeconomic flexibility. This contributed to an unusual degree of financial success and of commercial and geographical continuity across the trade. It pays to remember that early modern instruments were never in fact purely 'scientific' in the modern sense. They also belonged firmly to the world which, as Robert Dodsley wrote, was "a great Toy-shop, and all its Inhabitants run mad for Rattles".[56]

---

53 *Old Whig or The Consistent Protestant*, 17 November 1737.

54 *London Evening Post*, 10 November 1737; *Daily Post*, 12 November 1737; *Daily Gazetteer*, 12 November 1737.

55 Roger Lonsdale, *Eighteenth-Century Women Poets*, Oxford, 1989, p. 162.

56 Dodsley, op. cit. (n. 1), p. 27.

CHAPTER 2

Selling by the Book: British Scientific Trade Literature after 1800

Joshua Nall and Liba Taub

Introduction

In 1951 the businessman and scientific instrument collector Robert Stewart Whipple published *John Yarwell, or the Story of a Trade Card*. The paper, a lively mix of history of science, antiquarianism, and speculation about Isaac Newton, centres on the scant evidence presented by an apparently unique 1683 Yarwell trade card acquired by Whipple in 1937. The card had been used by Newton "for scribbling paper", hence its unlikely survival, and Whipple's analysis of it jumps between a translation and study of Newton's Latin notes recorded on its reverse, and the enticing images of Yarwell's optical instrument manufactures provided on its obverse (*Pl. 3*). As Whipple noted, the extreme rarity of surviving Yarwell instrumentation makes the trade card an important historical resource, offering a rare window onto the diverse range of items produced by and/or available from one of London's few preeminent seventeenth-century opticians.[1]

Many subsequent historians of scientific instrumentation and its trade have faced the same problem of sparse source material, against which trade cards have proved to be invaluable aids. As Michael Crawforth has shown, behind their elaborate designs and meticulous illustrations lies important information about instrument design, industry trends, commercial demands, makers' techniques and specialisms, and even clues to the instrument business's social context.[2] For curators, trade cards help pinpoint places and dates of object manufacture.[3] For historians of the trade, they are one resource amongst many

1 Robert S. Whipple, "John Yarwell, or the Story of a Trade Card", *Annals of Science* 7 (1951), pp. 62–69: 63. The British Museum holds a similar, later example of the Yarwell card.

2 Michael A. Crawforth, "Evidence from Trade Cards for the Scientific Instrument Industry", *Annals of Science* 42 (1985), pp. 453–554. See also: A. D. Morrison-Low, *Making Scientific Instruments in the Industrial Revolution*, Aldershot, 2007, pp. 213–229.

3 Henry R. Calvert, *Scientific Trade Cards in the Science Museum Collection*, London, 1971, pp. 7–9.

 | DOI 10.1163/9789004324930_003

that help trace the genealogy of a business or guild.[4] By revealing the advertising techniques employed by proprietors, trade cards also open up a window onto the commercial dialectic between maker and user/consumer, helping delineate the myriad and often subtle variations in instrument uses and markets.[5] It is not surprising, therefore, that descriptions and illustrations of trade cards are relied upon as a resource in many works on the history of early modern and Georgian scientific instrumentation.[6]

Yet in 1990 the authors of the *Handlist of Scientific Instrument-Makers' Trade Catalogues 1600–1914* were driven to lament that "many institutions [...] care little for this type of material, and many curators give it only casual attention." The ephemera of science was treated as trivial, they argued, and scholars too often ignored it as an historical source.[7] Despite the great gains made in the discipline of instrument studies over the preceding four decades, much of it dependent in part upon the information supplied by trade cards, ephemeral material appeared to be falling out of fashion, and with it out of favour with the archivists, librarians, and curators charged with its collection and preservation. Certainly, as the discipline grew historians of science in general turned their attention away from instrumentation and trade ephemera, in favour of

---

4 Gloria Clifton, *Directory of British Scientific Instrument Makers 1500–1851*, London, 1995, pp. xi–xv; M. A. Crawforth, "Instrument Makers in the London Guilds", *Annals of Science* 44 (1987), pp. 319–377.

5 D. J. Bryden, "Evidence from advertising for mathematical instrument making in London, 1556–1714", *Annals of Science* 49 (1992), pp. 301–336; D. J. Bryden and Dennis L. Simms, "Archimedes as an Advertising Symbol", *Technology and Culture* 34 (1993), pp. 387–391; D. J. Bryden and D. L. Simms, "Trade Ephemera, Archimedes and the Opticians of London", *Atti della Fondazione Giorgio Ronchi* 42 (2007), pp. 797–838; Roy Porter, Simon Schaffer, Jim Bennett and Olivia Brown, *Science and Profit in 18th-Century London*, Cambridge, 1985; John R. Millburn, *Retailer of the Sciences: Benjamin Martin's Scientific Instrument Catalogues, 1756–82*, London, 1986; Paolo Brenni, "19th Century Scientific Instrument Advertising", *Nuncius* 17 (2002), pp. 497–513; Alexi Baker, "'Precision,' 'Perfection,' and the Reality of British Scientific Instruments on the Move During the 18th Century", *Material Culture Review* 74–75 (2012), pp. 14–29: 17–19.

6 A useful overview of this literature, focused on the British scene, is given in Morrison-Low, op. cit. (n. 2), pp. 6–12. A good recent example of the use of trade cards as sources is: Brian Gee, Anita McConnell and A. D. Morrison-Low (eds.), *Francis Watkins and the Dollond Telescope Patent Controversy*, Aldershot, 2014, which reproduces eleven trade cards and bill heads of various London makers.

7 R. G. W. Anderson, J. Burnett and B. Gee, *Handlist of Scientific Instrument-Makers' Trade Catalogues 1600–1914*, Edinburgh, 1990, pp. i–vii.

manuscript, book, and periodical sources.[8] And for those relatively few who did remain focussed on the instrument trade, a shift in emphasis towards the modern era of science's material culture brought with it a more varied and copious supply of sources to pick through. Even once rich veins of alternative archival and bibliographic materials were recognised as ripe for investigation, the trade card's successor, the instrument manufacturer's trade catalogue, received little attention and credit as a potentially valuable source.[9] Whilst beautiful and always-rare early-modern trade cards had been carefully preserved and meticulously studied, seemingly cheap and evanescent post-1800 instrument catalogues were found by the *Handlist*'s authors to be uncatalogued, ignored, disintegrating, and even lost.[10]

Ironically, in the very same year as these authors' polemic, a seminal study by Deborah Jean Warner would demonstrate quite indubitably the importance of modern instrument trade literature as an historical source. As a curator in charge of acquiring and stewarding a diverse and amorphous collection, Warner was prompted to ask: "What is a scientific instrument, when did it become one, and why?" To tackle the question, she confronted the problem of terminology, offering a history of the use of the phrase 'scientific instrument.' Historians' deployment of this term had been widely noted as potentially anachronistic, as it had not been widely adopted until the mid-nineteenth century. Even then, and since that time, its application has been neither uniform nor unanimously agreed. To trace the term's complex genealogy, Warner provides evidence from the realm of manufacturing, through analysis of scientific sales and trade literature. So, for example, in France in 1787 Étienne Lenoir, ingénieur du Roi, was described as constructing "les instrumens à l'usage des sciences", and by 1853 the sales catalogue of Lerebours et Secretan referred to

---

8 John L. Heilbron, "Some Uses for Catalogues of Old Scientific Instruments", in R. G. W. Anderson, J. A. Bennett and W. F. Ryan (eds.), *Making Instruments Count: Essays on Historical Scientific Instruments presented to Gerard L'Estrange Turner*, Aldershot, 1993, pp. 1–16: 1–4.

9 There are, of course, some exceptions: John A. Chaldecott, "Printed Ephemera of Some 19th-Century Instrument Makers", in Christine Blondel, Françoise Parot, Anthony Turner and Mari Williams (eds.), *Studies in the History of Scientific Instruments: Papers Presented at the 7th Symposium of the Scientific Instrument Commission*, London, 1989, pp. 159–168; P. Brenni, "The Illustrated Catalogues of Scientific Instrument-Makers", in Ibid., pp. 169–178.

10 The internet has been an invaluable resource for reversing this trend, with large numbers of post-1800 instrument catalogues now available to view and download online. The Scientific Instrument Commission maintains a list of such websites: http://www.sic.iuhps.org/references/onlinecatalogues.shtml (accessed 19 February 2016).

"instruments de la science modern". The term 'Wissenschaftliche Instrument' probably came into use in Germany in the 1830s, and likely was well established by 1850 when two German immigrants to New York described themselves as "scientific instrument makers". In 1865 the London firm of R. and J. Beck published *An Illustrated Catalogue of Scientific Instruments*, and at the International Exhibition held at London in 1862 the Austrians stated that "the production of Scientific Instruments of every description has established its chief seat at Vienna". By the 1880s the prolific Philadelphia dealer J. and W. Queen could boast in one of their many catalogues that "The character and uses of scientific instruments are so varied and the stock so large as to require, in our establishment, their division into special departments".[11] In these examples, Warner demonstrates the singular importance of 'ephemeral' sales and trade literature as a resource for revealing and untangling the complex development of a discernably modern and global 'scientific instrument' industry.

The present essay aims to reinforce and extend the case made for the importance of modern scientific trade literature by the *Handlist*'s authors and by Warner. Here we consider ephemeral trade and sales material issued by the British trade in the Victorian era through to the early twentieth century. For our purposes such material encapsulates a diverse array of published formats and types, ranging from single-sheet advertising pamphlets through to lavishly illustrated hard-bound sales catalogues. We present here case studies drawn from our own and others' work, which collectively illustrate the various ways in which such literature can be used as a crucial resource for the historian, whether the subject of study is scientific instruments; the business of science; cultures of trade, commerce, and consumerism; Victorian expositions, shows, and shops; or the sciences' material practices. The basis for all our examples has been work carried out in the Whipple Museum of the History of Science, the institution that Robert Stewart Whipple founded when he donated his personal collection—including the Yarwell trade card—to the University of Cambridge in 1944.[12] The Whipple Museum has an extensive and growing collection of scientific trade literature, predominantly but not exclusively the products of British firms, and largely dating to the nineteenth and twentieth

11 Deborah Jean Warner, "What is a scientific instrument, when did it become one, and why?", *British Journal for the History of Science* 23 (1990), pp. 83–93: 86–88.

12 On the Museum's foundation, see: Liba Taub and Francis Willmoth (eds.), *The Whipple Museum of the History of Science: Instruments and Interpretations, to Celebrate the 60th Anniversary of R. S. Whipple's Gift to the University of Cambridge*, Cambridge, 2006, pp. 1–76; J. A. Bennett, "Museums and the History of Science at Oxford and Cambridge", *British Journal for the History of Science* 30 (1997), pp. 29–46.

centuries. As an active part of the University's Department of History and Philosophy of Science, the Museum has worked to make this valued collection available to both museum staff and researchers more broadly, including students. We are particularly grateful to be able to draw on work carried out in the Whipple by Kemal de Soysa, Boris Jardine, and Natalie Vokes. Taken as a whole, the case studies presented here demonstrate the rich variety of historical facts, clues, and lessons that can be drawn from an active engagement with the 'mere' ephemera of past science.

Display and Representation in the Age of Exhibitions

The nineteenth century was an era of extraordinary growth in the field of scientific instrument manufacture and use. The variety, quantity, and distribution of instruments expanded commensurate with the growth of scientific practices themselves.[13] Robert Anderson's valuable introductory survey to these trends highlights six broad categories of organisational, disciplinary, and social change in nineteenth-century Western science, each of which contributed crucially to this growth. Expansion of public education demanded new tools for mass instruction. The rise of laboratories as the vital site of scientific research compelled the development of more and new bench instrumentation. Growing professionalization lead to growing institutionalisation, with the establishment of numerous disciplinary societies, many of which founded their own collections, cabinets, and museums. Industrialisation necessitated scrutiny of processes and outputs. Colonialization boosted the need for accurate navigational and surveying instrumentation, as well as tools of imperial surveillance and control. And the establishment of public health legislation brought with it the formation of government and industrial testing laboratories.[14] One could perhaps add to Anderson's list the growth of a consumerist middle class, who themselves began to acquire scientific instruments as philosophical 'toys' (think of the wide variety of Victorian optical illusion devices that survive in science museum collections), or as hobby appliances

13 Useful general introductions are: Peter R. de Clercq (ed.), *Nineteenth-Century Scientific Instruments and their Makers: Papers Presented at the Fourth Scientific Instrument Symposium, Amsterdam 23–26 October 1984*, Amsterdam, 1985; Gerard L'E. Turner, *Nineteenth-Century Scientific Instruments*, London, 1983.

14 R. G. W. Anderson, "Were Scientific Instruments in the Nineteenth Century Different? Some Initial Considerations", in de Clercq (ed.), op. cit. (n. 13), pp. 1–12.

(think of the numerous desktop telescopes that were designed for home star-gazing and comet-hunting).[15]

The tools of science themselves not only changed, then, but also the markets for these tools. A history of scientific instruments in the nineteenth century is by necessity a history of business, of engineered technology, of labour and industry, of global commerce, of imperial control, of state bureaucracy, of mass education, of leisure and consumerism, of laboratory practice, and of the cultures that sustained these dynamic enterprises.[16] It is here that scientific trade literature can prove to be a vital historical source. Just like the venerated early-modern trade card, the ephemeral sales, trade, and advertising literature of the nineteenth-century instrument industry offers crucial insights into not only the hard tacks of commerce—who made what, where, in what quantities, and for how much did they sell?—but also the broader socio-economic trends that shaped and were shaped by an emergent techno-scientific industry.

Anderson presents one striking example of this, in his account of the marked depersonalisation of the instrument trade over the century. Trade catalogues are the crucial source here, as they attest to a transition from products issued under the name of an individual craftsman to those advertised under the auspices of a commercial enterprise, some of which, like the Cambridge Scientific Instrument Company (founded 1881), operated as a 'limited company' and incorporated no personal names into their title.[17] This transition from

15 G. L'E. Turner, "Scientific Toys", *British Journal for the History of Science* 20 (1987), pp. 377–398: 383–398; Melanie Keene, *Object Lessons: Sensory Science Education, 1830–1870*, PhD dissertation, University of Cambridge, 2008.

16 Bernward Joerges and Terry Shinn (eds.), *Instrumentation Between Science, State and Industry*, Dordrecht, 2001; M. E. W. Williams, *The Precision Makers: A History of the Instrument Industry in Britain and France, 1870–1939*, London, 1994; Michael J. G. Cattermole and Arthur F. Wolfe, *Horace Darwin's Shop: A History of the Cambridge Scientific Instrument Company 1878–1968*, Bristol, 1987; Robert Bud and Susan E. Cozzens (eds.), *Invisible Connections: Instruments, Institutions, and Science*, Bellingham, 1992; Iwan Rhys Morus, "Manufacturing Nature: Science, Technology and Victorian Consumer Culture", *British Journal for the History of Science* 29 (1996), pp. 403–434; Bruce J. Hunt, "The Ohm Is Where the Art Is: British Telegraph Engineers and the Development of Electrical Standards", *Osiris* 9 (1994), pp. 48–63; Graeme J. N. Gooday, *The Morals of Measurement: Accuracy, Irony, and Trust in Late Victorian Electrical Practice*, Cambridge, 2004, pp. 1–39; M. Norton Wise (ed.), *The Values of Precision*, Princeton, 1995; Ben Marsden and Crosbie Smith, *Engineering Empires: A Cultural History of Technology in Nineteenth-Century Britain*, Basingstoke, 2005; Daniel R. Headrick, *The Tools of Empire: Technology and European Imperialism in the Nineteenth Century*, New York, 1981.

17 Anderson, "Were Scientific Instruments ...", op. cit. (n. 14), pp. 2–3. Cambridge Instruments were certainly trail blazers with respect to eschewing personal names in their title; the practice became more common in the early twentieth century.

identification with a particular person to that of a seemingly anonymous entity might be characterised as the concomitant of a much more profound shift in the industry of science, namely the displacement of artisanal skills by industrial manufacture. Just as much as with nomenclature, we can also follow this transformation through attention to the visual aesthetic of trade literature. The iconography of early modern trade cards, with their focus on the individual maker and his wares, gives way during the nineteenth century to catalogues devoid of people, but festooned with elaborate illustrations and diagrams of factories, workshops, and complex instrumental setups (*Fig. 2.1*).[18]

As Kemal de Soysa has noted in a study of Victorian trade literature in the Whipple Museum's possession, of particular interest in regard to these transformations is the "methods used by makers to establish their status in the market".[19] Depersonalisation risked further blurring the already slippery distinction between makers and retailers, and with it obscuring the unique qualities of individual manufacturers. Making a growing and diversifying consumer base aware of the diversifying range of instruments produced, amongst an expanded field of instrument manufacturers, presented new challenges to those active in the market.

One of the most significant responses to these changes was the rise of a global culture of exhibitions. As Robert Brain has put it:

> During the nineteenth century "going to the exhibition" became an important pastime for Europeans of all classes. International Exhibitions [...] came to regulate the rhythms of production, commerce, and leisure for nineteenth century Europeans [...].[20]

Conceived as public events that could amalgamate popular education, commercialism, and international competition, the London Great Exhibition of 1851 and the many that followed it aspired to a panoramic vision of modernity.

---

18 Brenni, "Illustrated Catalogues...", op. cit. (n. 9); Willem D. Hackmann, "The Nineteenth-Century Trade in Natural Philosophy Instruments in Britain", in de Clercq (ed.), op. cit. (n. 13), pp. 53–91: 69–72; Williams, *The Precision Makers*..., op. cit. (n. 16). The latter two works caution that in the case of the British instruments industry the rise of factory mechanisation and large business corporations was only a partial and late development, with some workshop and family firms surviving alongside newer, industrial enterprises into the twentieth century. Even when not associated with particular individuals, the brand identity of a company was a valuable asset.

19 Kemal de Soysa, *The Victorian Scientific Instrument-Maker's Trade Catalogue: Marketing and Patronage*, MPhil essay, University of Cambridge.

20 Robert Brain, "Going to the Exhibition", in Richard Staley (ed.), *The Physics of Empire*, Cambridge, 1994, pp. 113–142: 113.

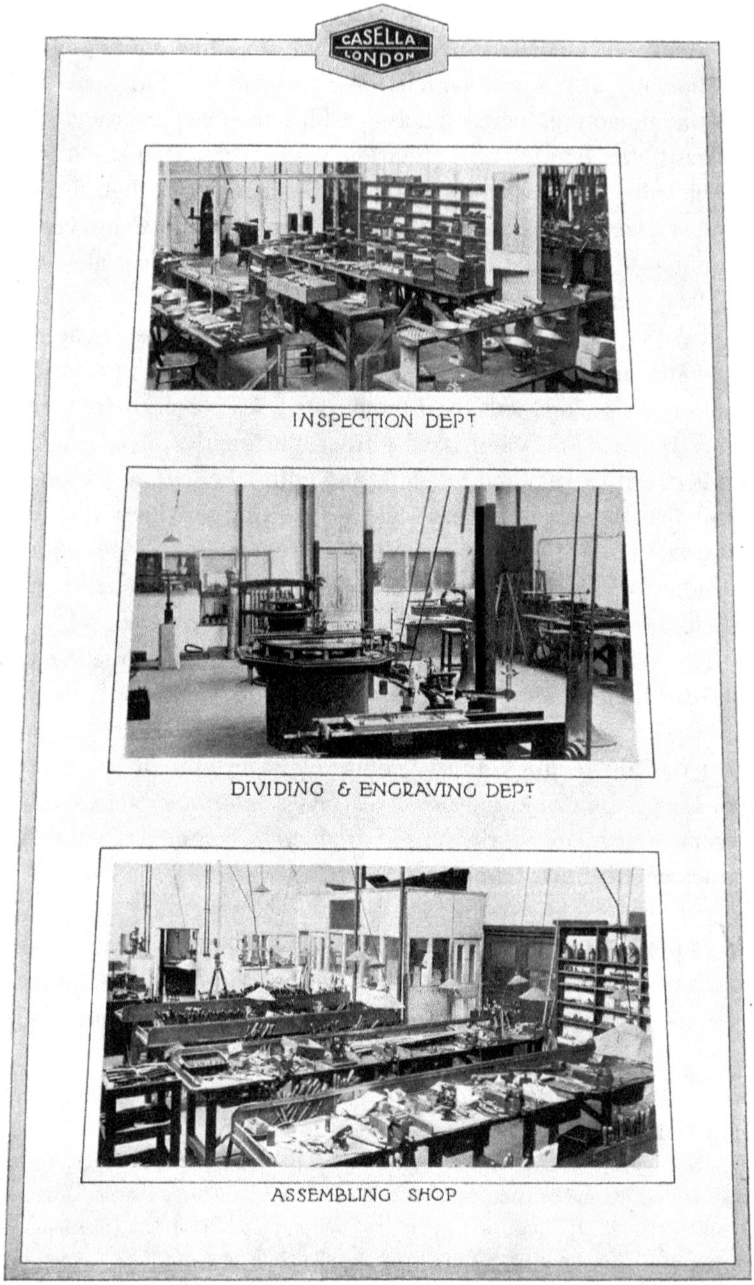

FIGURE 2.1 *Like many instrument-making firms in the industrial age, Casella used photographs of their factory floor and workshops as frontispieces to their catalogues. From* Surveying and Drawing Instruments, *1923.*
IMAGE © WHIPPLE MUSEUM, CAS.15.

At the heart of this new exhibition culture was a creed of public display of—and competitive judgement upon—cutting-edge manufactures. In the introduction to the official catalogue of the 1851 Exhibition, Henry Cole proclaimed that "The activity of the present day chiefly develops itself in commercial industry, and it is in accordance with the spirit of the age that the nations of the world have now collected together their choicest productions".[21] In an increasingly standardised international market, the tradition of concealing trade secrets gave way to a new vogue for public scrutiny and open comparison of rival manufacturers' wares. Vast quantities of commercial products were collected, catalogued, and displayed, deliberately exposing the mechanisms of industrial knowledge and technique.[22]

This new form of 'exhibition representation' soon came to be regarded by the manufacturers themselves as a productive and essential component of their identities. "Firms came to regard the representation of their products", Brain suggests, "as equal in importance to the products' quality".[23] It is not surprising, therefore, that the era of exhibition culture coincides with a flourishing of lavishly illustrated trade catalogues, pamphlets, and technological handbooks. As exhibitions themselves were relatively infrequent, forms of exhibition display and aggrandizement were best disseminated in mass-produced sales literature. Mobile ephemera was a powerful tool for extending exhibition representation beyond the walls of the exposition hall and into the laboratories and homes of consumers; indeed, catalogues themselves were an important element of the growing culture of aspirational consumerism in the latter part of the nineteenth century.

A beautifully illustrated *c.* 1891 catalogue by Negretti and Zambra in the Whipple Museum's collection is exemplary of this wider trend. Just as many of its rivals did, the firm placed the medals they had been awarded at international exhibitions on the page after the catalogue's frontispiece, highlighting the importance of exhibition participation and success to their reputation. Pride of place was given to the medal won at the Great Exhibition of 1851, awarded only a year after the firm had been founded. Across their trade literature Negretti and Zambra proudly proclaimed themselves the only recipients of a Prize Medal for meteorological instruments at the London exhibition,

21 *Official Descriptive and Illustrated Catalogue of the Great Exhibition of the Work of All Nations*, London, 1851, v. 1, pt. 1, p. 1; quoted in Hackmann, op. cit. (n. 18), p. 62.

22 Brain, "Going to the Exhibition", op. cit. (n. 20); R. M. Brain, *Going to the Fair: Readings in the Culture of Nineteenth-Century Exhibitions*, Cambridge, 1993, pp. 9–18; Tony Bennett, "The Exhibitionary Complex", *New Formations* 4 (1988), pp. 73–102.

23 Brain, "Going to the Exhibition", op. cit. (n. 20), p. 119.

conveniently failing to mention that four Council Medals—a higher honour—were awarded to other manufacturers (*Fig. 2.2*).[24]

Lists of the medals won at exhibitions were augmented by notices of Negretti and Zambra's royal warrants, as well as the significant institutions which patronised the firm, such as the Board of Trade and the Royal Observatory. The greatest space and labour, however, was apportioned to lavish illustrations of the sites of exhibition and sale, as well as of the objects themselves. Visualising the elaborate displays presented at places like the Crystal Palace at Sydenham, as well as showing off the grand shops operated by the firm in trendy London streets, reinforces the central importance of representation to the post 1850s consumer culture in which firms like Negretti and Zambra operated. Ostentatious display substantiated the firm's commercial bona fides, and elaborate technical diagrams of the objects for sale reinforced the firm's expertise and skill. Trade literature is, therefore, an important but understudied example of how historical actors established their credibility and trustworthiness through carefully-managed display, exhibition, and self-presentation. As Iwan Rhys Morus and others have shown, the commercial world of Victorian science compelled individuals and firms to become showmen, with public representations of themselves and their artefacts "the outcome, more often than not, of careful stage management".[25] The transition from the closed world of 'gentleman science' and secretive artisanal production to the public world of exhibition culture needs to be understood in part, then, as a change in the commercial philosophy of display and advertisement more broadly throughout nineteenth-century culture, and not only in Britain. What to show, and what not to show, was the crucial question, and trade literature provides a revealing map of how instrument manufacturers changed their answers to this over time. Certainly by 1891, Negretti and Zambra, like most of their rivals, were not shy about exposing the prime technical details of their instruments (*Fig. 2.3*). In the era of expositions, it paid to build trust through open display.

---

24 De Soysa, op. cit. (n. 19).

25 I. R. Morus, "Working Out in the Nineteenth Century", *Studies in History and Philosophy of Science* 38 (2007), pp. 605–609: 608. See also: Morus, "Manufacturing Nature…", op. cit. (n. 16); Marsden and Smith, op. cit. (n. 16), pp. 4–11; Aileen Fyfe and Bernard Lightman (eds.), *Science in the Marketplace: Nineteenth-Century Sites and Experiences*, Chicago, 2007; Joe Kember, John Plunkett and Jill A. Sullivan (eds.), *Popular Exhibitions, Science and Showmanship, 1840–1910*, London, 2012.

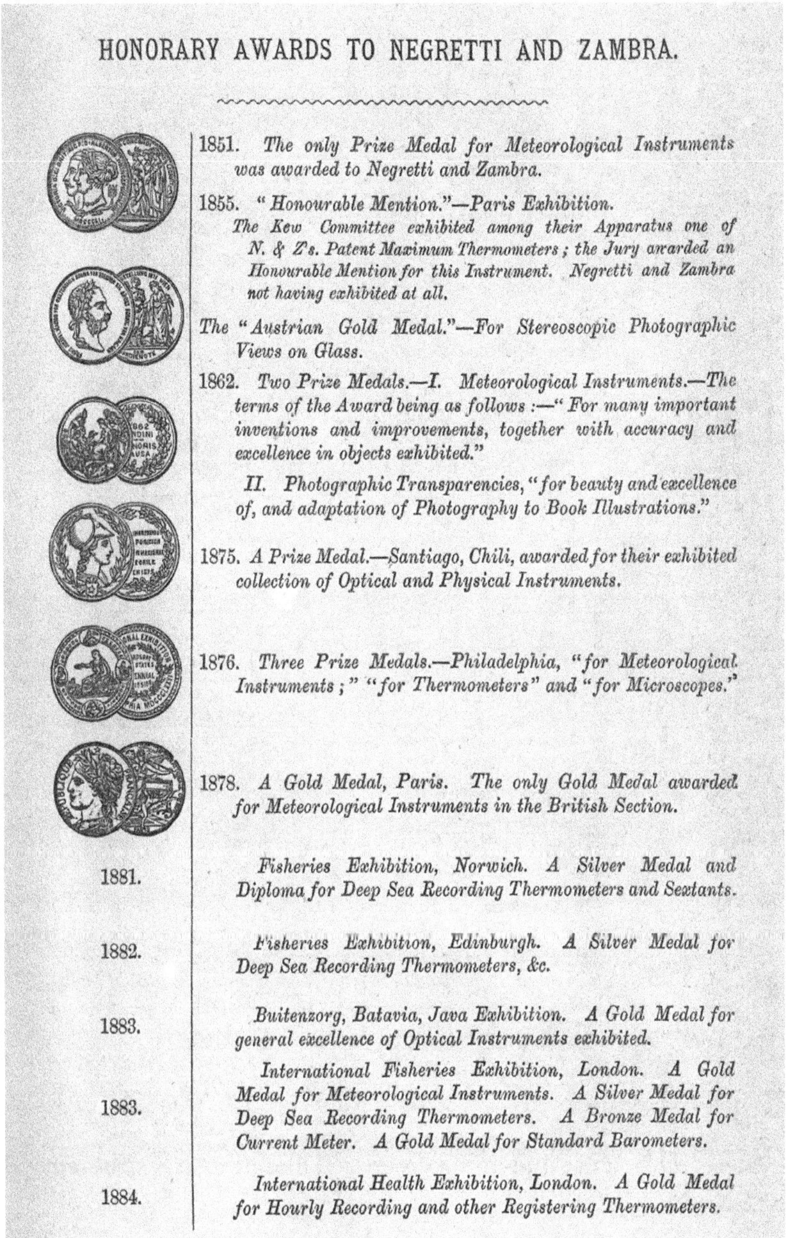

HONORARY AWARDS TO NEGRETTI AND ZAMBRA.

1851. *The only Prize Medal for Meteorological Instruments was awarded to Negretti and Zambra.*

1855. "*Honourable Mention.*"—*Paris Exhibition.*

The Kew Committee exhibited among their Apparatus one of N. & Z's. Patent Maximum Thermometers; the Jury awarded an Honourable Mention for this Instrument. Negretti and Zambra not having exhibited at all.

The "Austrian Gold Medal."—For Stereoscopic Photographic Views on Glass.

1862. *Two Prize Medals.—I. Meteorological Instruments.—The terms of the Award being as follows:—" For many important inventions and improvements, together with accuracy and excellence in objects exhibited."*

II. Photographic Transparencies, "for beauty and excellence of, and adaptation of Photography to Book Illustrations."

1875. *A Prize Medal.—Santiago, Chili, awarded for their exhibited collection of Optical and Physical Instruments.*

1876. *Three Prize Medals.—Philadelphia, "for Meteorological Instruments;" "for Thermometers" and "for Microscopes."*

1878. *A Gold Medal, Paris. The only Gold Medal awarded for Meteorological Instruments in the British Section.*

1881. *Fisheries Exhibition, Norwich. A Silver Medal and Diploma for Deep Sea Recording Thermometers and Sextants.*

1882. *Fisheries Exhibition, Edinburgh. A Silver Medal for Deep Sea Recording Thermometers, &c.*

1883. *Buitenzorg, Batavia, Java Exhibition. A Gold Medal for general excellence of Optical Instruments exhibited.*

1883. *International Fisheries Exhibition, London. A Gold Medal for Meteorological Instruments. A Silver Medal for Deep Sea Recording Thermometers. A Bronze Medal for Current Meter. A Gold Medal for Standard Barometers.*

1884. *International Health Exhibition, London. A Gold Medal for Hourly Recording and other Registering Thermometers.*

FIGURE 2.2 *Negretti and Zambra followed the common nineteenth-century practice of listing at the front of their catalogues the various jury medals they had been awarded at international exhibitions.*

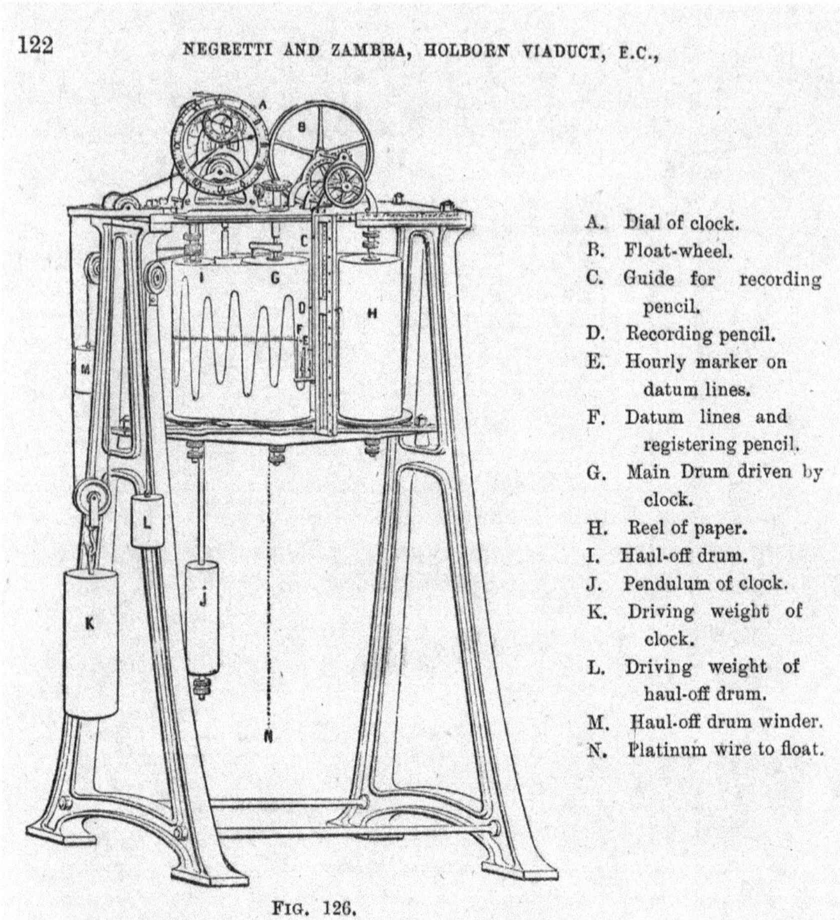
122 NEGRETTI AND ZAMBRA, HOLBORN VIADUCT, E.C.,

A. Dial of clock.
B. Float-wheel.
C. Guide for recording pencil.
D. Recording pencil.
E. Hourly marker on datum lines.
F. Datum lines and registering pencil.
G. Main Drum driven by clock.
H. Reel of paper.
I. Haul-off drum.
J. Pendulum of clock.
K. Driving weight of clock.
L. Driving weight of haul-off drum.
M. Haul-off drum winder.
N. Platinum wire to float.

FIG. 126.

FIGURE 2.3 *Rather than keep novel designs secret, Victorian firms like Negretti and Zambra mirrored trends in public exposition by revealing the technical details of their products through their trade catalogues, as in this labelled diagram of a tide-gauge.* IMAGE © WHIPPLE MUSEUM, NEG.4.

Research Science and Commercial Collaboration in Instrument Design

There was an obvious danger to the way firms such as Negretti and Zambra advertised themselves in the manufacturing age. By positioning themselves at the forefront of the public, commercial world exemplified by grand expositions, these firms were at a risk of simultaneously distancing themselves from the field of 'pure' research science. When Henry Sutton died of the plague in

London in 1665, founder-member of the Royal Society Robert Moray lamented that "we all here [in Oxford] are much troubled with the loss of poor [...] Sutton".[26] In the early-modern world of natural philosophy the relationship between the practitioner and the instrument maker was often very close. Indeed, in the case of someone like John Bird, the boundary between the two blurs, given his very close working relationship with a number of astronomers. Conversely, the large-scale manufacture of instruments in the nineteenth century by firms like Negretti and Zambra placed distance between the industrial world of often mass-produced instruments and the research world of scientists. In so doing, they risked appearing less like scientific instrument makers and more like commercial retailers of domestic products.

This problem was exacerbated by a growing homogenisation of the instrument market. Paolo Brenni has observed that the commonplace recycling of xylographic engravings across the industry's catalogues demonstrates a pervasive habit of firms either directly to copy the manufactures of their rivals, or resell the wares of others under their own name.[27] The consequence was a notable enlargement of the typical firm's catalogue—many in the Whipple's collection are the size of a large 300+ page hardback book—but also a consequent convergence of their inventory.

For firms jostling to establish and enhance their status within a crowded and increasingly undifferentiated market, closing the gap between industry and the working worlds of scientists therefore became an important form of promotion. For many firms, the most impressive way of doing this was to directly link their products to a recognised name in the world of science. Establishing a connection between a particular instrument and a specific researcher could single out an item as unique, and it could imbue it with the cachet of a famous name. As a technique of advertising, this was by no means a new ploy. Instrument makers since the early modern period had worked to link their wares with names such as Archimedes, Newton, and Boyle.[28] What is distinct to the emerging modern market of the nineteenth and early twentieth

---

26 Robert Moray to Henry Oldenburg, 10 October 1665, in A. Rupert Hall and Marie Boas Hall (eds.), *The Correspondence of Henry Oldenburg*, London and Madison, WI, 1965–86, 13 vols.: v. 2, p. 561, quoted in: J. A. Bennett, "Early Mathematical Instruments", paper given at *A Symposium to Mark the 350th Anniversary of The Royal Society*, Wadham College, Oxford, 25 November 2010.

27 Brenni, "Illustrated Catalogues ...", op. cit. (n. 9).

28 Jeffrey R. Wigelsworth, *Selling Science in the Age of Newton: Advertising and the Commoditization of Knowledge*, Farnham, 2010, pp. 147–174; Bryden and Simms, "Archimedes ...", op. cit. (n. 5); Bryden and Simms, "Trade Ephemera ...", op. cit. (n. 5).

century, however, is the working relationships of those involved. For the most part early-modern makers either traded on the cachet of their own name (such as in the cases of Sutton and Bird), or they simply used (one might say stole) the names and likenesses of famous natural philosophers or mathematicians in an attempt to feed off the prestige of these celebrity brands. Manufacturing firms from the Victorian era onwards, in contrast, pursued links with recognised men of science by forging working relationships with them. Such collaborations could produce much more than just brand-name recognition; significantly, practical partnerships between firms and working scientists helped shape the design, development, and promotion of a wide variety of instruments.

The trade literature of the microscope makers Smith and Beck provides one illuminating example. Boris Jardine, in his study of Charles Darwin's microscopy, has shown how Darwin sought to "actively establish for himself a role within the scientific elite," one key component of this move being a demonstration of his skill as a microscopist.[29] On the one hand, this involved working with a state-of-the-art achromatic compound desk microscope, purchased from Smith and Beck in 1847 for the then considerable sum of £ 34. But on the other hand, it also involved demonstrating a combined mastery of the standard field microscope of the day, the portable single-lens 'simple' microscope. Darwin had taken a simple aquatic microscope by Banks on the Beagle voyage, for which he had developed "quite a hatred" owing to a number of flaws he perceived in its composition.[30] At around the same time that he purchased his compound microscope, Darwin therefore also engaged Smith and Beck to make for him a replacement single-lens instrument to his own design. In a letter to Richard Owen, Darwin wrote that this new type had so pleased the microscope makers that "they have made another as a model" to display in their shop.[31] Darwin's contribution to John Herschel's manual of scientific enquiry for the British navy, *On the use of the microscope on board ship*, described this instrument in detail and explicitly informed readers that "A microscope such as here described, and most of the apparatus, can be seen at Messrs. Smith & Beck's, opticians, of Coleman Street, London".[32]

29 Boris Jardine, "Between the Beagle and the Barnacle: Darwin's Microscopy, 1837–1854", *Studies in History and Philosophy of Science* 40 (2009), pp. 382–395: 394.

30 Charles Darwin to Richard Owen, 26 March 1848, in: Frederick Burkhardt et al. (eds.), *The Correspondence of Charles Darwin*, Cambridge, 1985-ongoing, 22 vols.; v. 4, p. 125.

31 Ibidem.

32 Charles Darwin, "On the Use of the Microscope on Board Ship", in John F. W. Herschel (ed.), *A Manual of Scientific Enquiry; Prepared for the Use of Her Majesty's Navy; and Adapted for Travellers in General*, London, 1849, 389–395: p. 393.

Thus began a symbiotic relationship of collaboration and promotion. Darwin got to attach his name to a novel, specialist microscope design, bolstering his own credentials as a pre-eminent expert in field microscopy. And Smith and Beck began to incorporate 'Mr Darwin's Single Microscope' into their trade and publicity material, linking their firm to the work of a famous man of science. In 1852 John Quekett's *Practical Treatise on the Use of the Microscope* puffed the instrument as having "proved a very useful working tool in the hands of Mr. Darwin".[33] Over a decade later, the instrument still featured in a prime example of a form of trade literature typical of the era, the technical-manual-cum-sales-catalogue. Richard Beck's *Treatise on the Construction, Proper Use, and Capabilities of Smith, Beck, and Beck's Achromatic Microscopes* included an illustrated section on "Darwin's Instrument", promoting it as "made to the express design of Mr. Darwin" and boasting that "it is with such arrangements that he has carried on many of his valuable investigations" (*Fig. 2.4*).[34] Jardine notes that no extant examples of this model of microscope are known, this in itself suggesting that "it was the presence of Darwin's name in their catalogues, rather than prolific sales of the microscope itself, that benefited the firm".[35]

In other cases, similar forms of collaboration could have a profound impact on the development and success of an entire class of instruments. The Whipple Museum's vast collection of over two thousand pieces of sales and trade literature issued by the Cambridge Instrument Company (CIC) is a revealing source in this regard.[36] Effectively covering the entire history of the firm, the Whipple's collection offers a remarkable source for tracing both the genealogy and biography of a specific object or class of instruments. Nathalie Vokes, in attempting to understand the apparently conservative development of a portable electrocardiograph (ECG) by the firm during the 1920s and 1930s, found crucial evidence in this trade literature. In scaling down their pioneering bench-sized ECG, CIC chose to stick with a cumbersome string galvanometer design rather than shifting to smaller and lighter—but less tried and tested—valve-based technology. Vokes explains this conservative design philosophy in

33 John Quekett, *A Practical Treatise on the Use of the Microscope*, London, 1852, p. 62.

34 Richard Beck, *A Treatise on the Construction, Proper Use, and Capabilities of Smith, Beck, and Beck's Achromatic Microscopes*, London, 1865, p. 102.

35 Jardine, op. cit. (n. 29), p. 388.

36 The firm was founded with the name Cambridge Scientific Instrument Company, subsequently changing its name slightly on a variety of occasions, as outlined in Cattermole and Wolfe, op. cit. (n. 16), p. xiv. For simplicity's sake it is referred to here as CIC throughout, the name it used from 1924 until 1967.

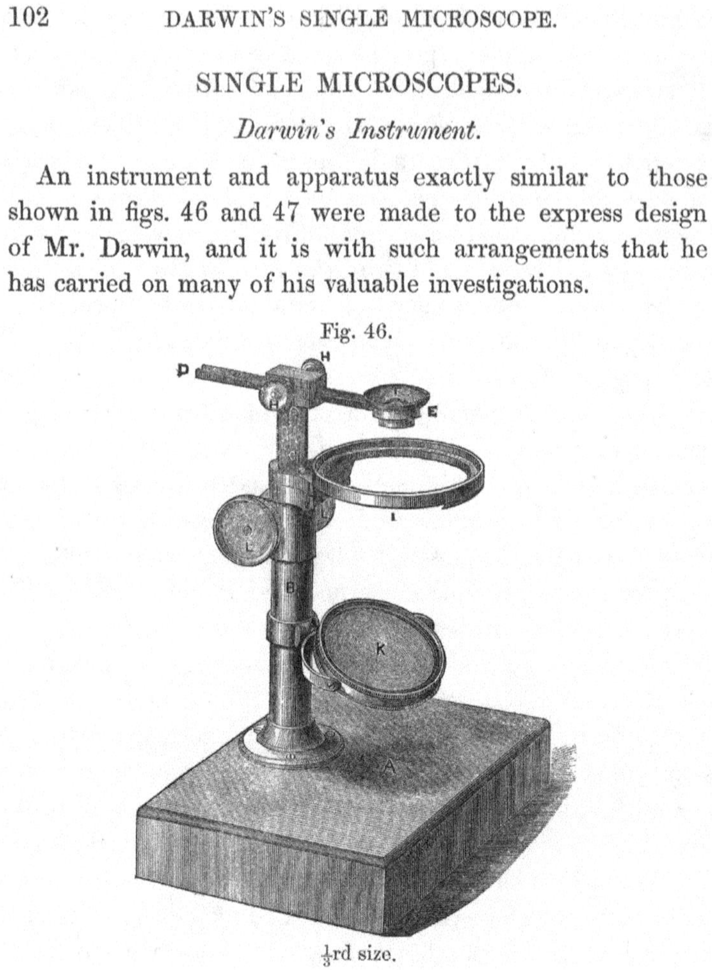
102 DARWIN'S SINGLE MICROSCOPE.

SINGLE MICROSCOPES.

Darwin's Instrument.

An instrument and apparatus exactly similar to those shown in figs. 46 and 47 were made to the express design of Mr. Darwin, and it is with such arrangements that he has carried on many of his valuable investigations.

Fig. 46.

$\frac{1}{3}$rd size.

FIGURE 2.4 *"Darwin's Instrument", featured in Richard Beck's 1865 manual and catalogue (no. 33), exemplifies the symbiotic relationships forged by a number of Victorian makers and eminent men of science.*
IMAGE © WHIPPLE MUSEUM, CR 5:30.

part by considering how the firm marketed its new design.[37] CIC were keen to exploit both their own status as founders of the field, as well as clinical practitioners' long-built familiarity with CIC's instruments. The firm's sales literature

37 Natalie Vokes, "Remodeling the Electrocardiograph at the Cambridge Scientific Instrument Company", MPhil essay, University of Cambridge. See also: Cattermole and Wolfe, op. cit. (n. 16), pp. 220–236.

from the era therefore repeatedly emphasised continuity of design and operation from the old bench to the new portable design. For example, an eight-page sales brochure from 1929 stressed that "In order that the records obtained with this new model may be strictly comparable with all published work to date, the recording system adopted is the same as that used in the apparatus employed by the workers in cardiology since the commencement of this method of diagnosis".[38]

This claim to continuity was crucial for maintaining the firm's close ties with a specific research scientist, the celebrated cardiologist Sir Thomas Lewis. Lewis had been a pioneer in the use of the ECG as a clinical tool, and crucially he had consulted with CIC in the early days of the instrument's development. As late as 1939, twenty-eight years after Lewis had collaborated on the development of the first 'Cambridge Electrocardiograph', CIC were still boasting in a glossy gatefold pamphlet (*Fig. 2.5*) that:

> [t]he pioneering work of Sir Thomas Lewis was carried out on one of the first Cambridge Electrocardiographs [...] and this instrument, after over 25 years' continuous service, is still in use [...] it is no exaggeration to claim that the progress in the Science of Electrocardiography and the development of the Cambridge Electrocardiograph have been simultaneous.[39]

Lewis, who maintained in his clinical work an insistence on the use of a string galvanometer ECGs, thus not only helped shape the firm's original instrument design, but all subsequent designs over the next half-century. The first CIC ECG to use a valve amplifier was released in 1952, seven years after Lewis's death.

Despite the apparently cautious instrument development that resulted from this collaboration, there can be little doubt that such partnerships were hugely important to CIC, and hugely influential in their early success. The firm made its name as a manufacturer of complex, often niche, instrumentation suitable for the research field, and it certainly gained a great deal of credibility from the active involvement of prominent scientists, many of them associated with the University of Cambridge. Within seven months of its foundation CIC had established a royalty agreement with Dr Charles Smart Roy, Professor of Pathology at the University, to manufacture 'Roy's freezing microtome'. Roy

---

38 "Cambridge Portable Electro-Cardiograph", Cambridge Instrument Company List No. 980 (1929), p. 2.

39 "Cambridge Electrocardiograph for Battery or A. C. Mains Operation", Cambridge Instrument Company Folder No. 69 (1939), p. 2.

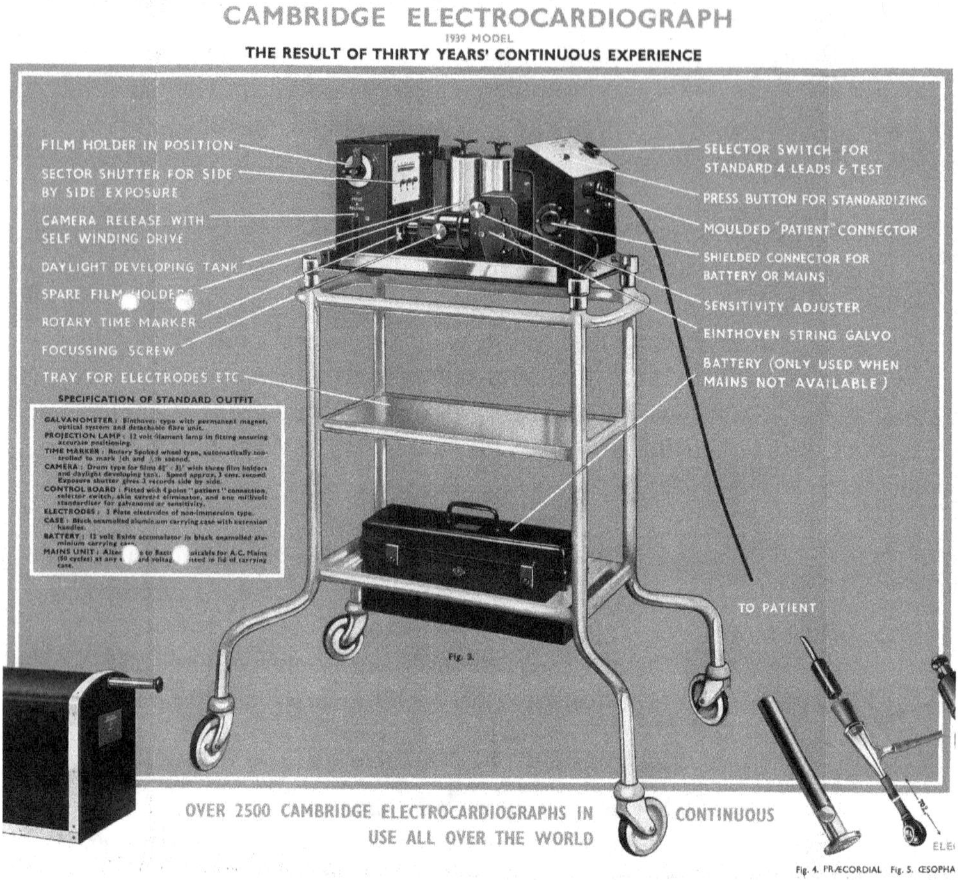

FIGURE 2.5 *This 1939 pamphlet issued by the Cambridge Instrument Company to promote their portable electrocardiograph stresses that their design is the "result of thirty years' continuous experience", providing a crucial insight into the Company's design philosophy and commercial strategy.*
IMAGE © WHIPPLE MUSEUM, CSI.261.

even agreed to personally test all the instruments of his design before they left the CIC's workshop. A decade later the Company's comprehensive descriptive catalogue listed no fewer than thirteen different instruments designed by and sold under Roy's name.[40] In the Cavendish Laboratory the Company assisted Lord Rayleigh with his metrological work on the British Association Standard

40 Cattermole and Wolfe, op. cit. (n. 16), pp. 24–26; "A Descriptive List of Instruments Manufactured and Sold by the Cambridge Scientific Instrument Company", List No. 12 (1891), p. 126.

Coil Apparatus, subsequently marketing the 'Rayleigh Electric Motor'. Later the firm would collaborate extensively with one of J. J. Thomson's acolytes, Hugh Callendar, advertising a wide range of temperature measuring and recording equipment under Callendar's name.[41] But it was in the rapidly growing discipline of experimental physiology that the CIC found its most fruitful partner in Keith Lucas. A Fellow of Trinity College from 1904 and one of the brightest young researchers in his field, Lucas studied muscle contraction and nerve action by re-engineering existing tools or inventing new instrumentation, pioneering a number of important physiological bench techniques in the process. At the age of twenty-seven Lucas joined the CIC Board of Directors, signing over to the Company patents he had already secured and actively assisting with design work on new instruments, all whilst remaining a University lecturer. Of the nine instrument types advertised in the 1913 CIC sales catalogue *Some Physiological Apparatus*, seven were listed as being wholly or partly designed by Lucas (*Fig. 2.6*).[42]

Two points that deserve further investigation are immediately evident from this brief foray into the Whipple Museum's collection of CIC sales and trade literature. Firstly, this ephemeral material provides an insight into the little-considered theme of research commercialisation in the era in which the laboratory sciences matured. Research patents, spin-off companies, and University-business start-ups are all enterprises much more likely to be associated with the twenty-first century university than its late nineteenth century predecessor. Yet in clear contradiction to recent claims that British academic scientists at the turn of the twentieth century "explicitly rejected private enterprise", trade and sales literature provide compelling evidence for rampant and aggressive commercialisation of the material culture of the era's cutting-edge research work.[43] Even after World War I, when expanded state support is supposed to have released academics from the final vestiges of entrepreneurial pursuits, we find the Cambridge University geologist Archibald T. J. Dollar undertaking the patenting and commercial exploitation of his novel

---

41 Cattermole and Wolfe, op. cit. (n. 16), pp. 27, 46–49, 185–190 and 202–206. "A Descriptive List…", op. cit. (n. 40), pp. 13 and 29. For Callendar equipment, see, for example: "Callendar Electric Recorders, Platinum Thermometers, and Apparatus for the Measurement of Small Resistances", Cambridge Scientific Instrument Company (1901).

42 Cattermole and Wolfe, op. cit. (n. 16), pp. 62–64; Robert G. Frank Jr., "Instruments, Nerve Action, and the All-or-None Principle", *Osiris* 9 (1994), pp. 208–235: 213–217. "Some Physiological Apparatus", Cambridge Scientific Instrument Company List No. 124 (1913).

43 Christine MacLeod, "Reluctant Entrepreneurs: Patents and State Patronage in New Technosciences, circa 1870–1930", *Isis* 103 (2012), pp. 328–339: 328.

The Cambridge Scientific Instrument Co., Ltd.,

THE LUCAS MUSCLE TROUGH.

FOR MUSCLE AND NERVE EXPERIMENTS.

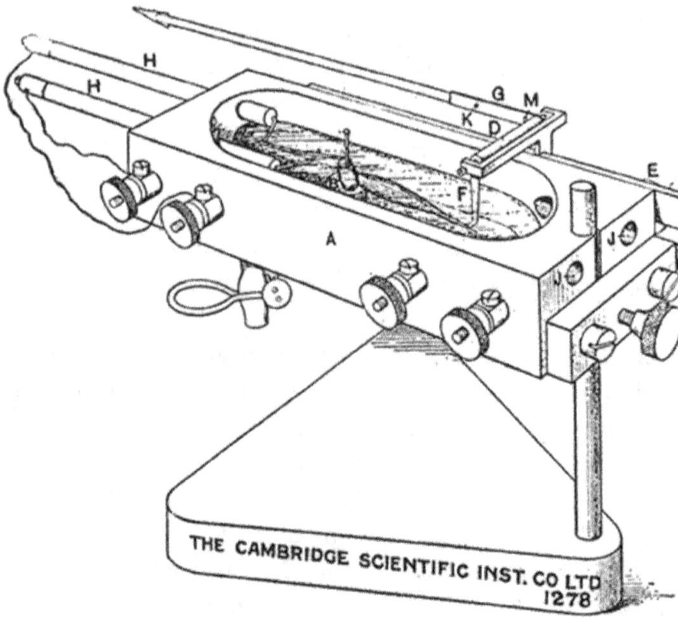

Fig. 12.

FIGURE 2.6 *The Cambridge Scientific Instrument Company released a wide range of instruments that took their name from the academic collaborators who worked with the firm, such as this muscle trough designed by Keith Lucas.*
IMAGE © WHIPPLE MUSEUM, CSI.150.

petrological integrating micrometer in partnership with one of CIC's own offshoots, Unicam Instruments (*Fig. 2.7*).[44]

Secondly, an understanding of CIC's Lucas partnership and others like it focuses attention on the complex transition that occurs when an instrument develops from a unique component within a novel experimental setup to a commercially-available piece of mass-produced bench equipment. As the

44 Archibald T. J. Dollar, "An Integrating Micrometer for the Geometrical Analysis of Rocks", *Mineralogical Magazine* 24 (1937), pp. 577–594; "Dr. Dollar's Integrating Micrometer", in untitled Unicam Instruments sales catalogue, c. 1940, section S.

UNICAM INSTRUMENTS
CAMBRIDGE

DR. DOLLAR'S INTEGRATING MICROMETER

(PAT. APPLIC. No. 29874)

FOR THE QUANTITATIVE ANALYSIS OF
TRANSPARENT OR OPAQUE SUBSTANCES
BY THE ROSIWAL METHOD

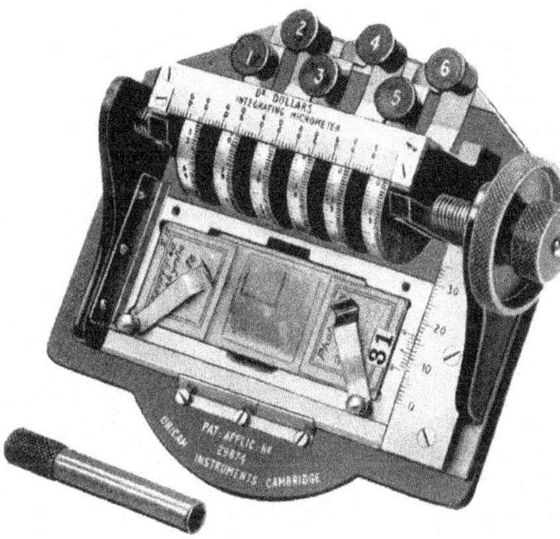

Fig. 1. Upper surface of Micrometer.

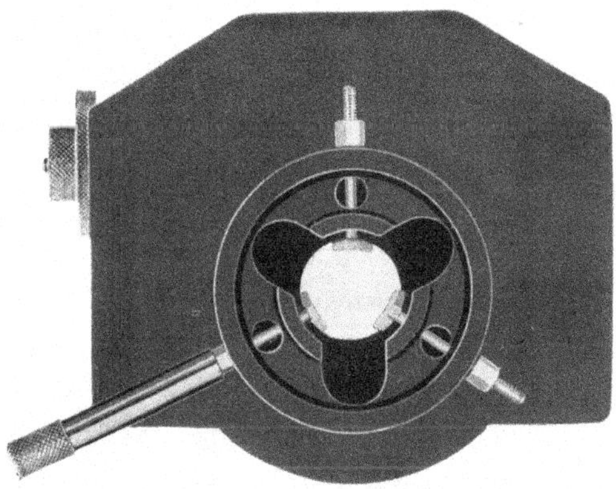

Fig. 2. Lower surface of Micrometer.

S.40

FIGURE 2.7 *Unicam Instruments was founded in 1934 by a former foreman of the Cambridge Instrument Company. Its trade literature is further evidence of fruitful partnerships between instrument making firms and research scientists such as Archibald T. J. Dollar.*
IMAGE © WHIPPLE MUSEUM, UNI.1.

discipline of experimental physiology grew, CIC were able to supply standardised pieces of equipment that embodied aspects of Lucas's skill as an experimental practitioner. Sales literature and instruction manuals produced to accompany these instruments help to explicate how the tacit knowledge essential for Lucas's own work was transformed into 'black box' apparatus suitable for use in a laboratory on a different continent to Lucas's own.[45] Though outside the remit of this present study, both these points deserve closer scrutiny, a job for which sales and trade literature will be an essential resource.

45 Heilbron, op. cit. (n. 8), pp. 10–13. On the 'black boxing' of scientific instruments see: Trevor Pinch, *Confronting Nature: The Sociology of Solar-Neutrino Detection*, Dordrecht, 1986, pp. 212–214.

Conclusion

Recently, historians of science have been paying closer attention to material that has often been regarded as somewhat ephemeral, for example newspapers, popular science magazines, and scientific toys. Scientific trade literature, often ignored as a resource for the historian of science, deserves similar special attention, and we hope to have shown here some examples of work that exemplify the success of using this kind of material. We have not attempted here to give you a complete overview of the work done by de Soysa, Jardine, Vokes, and ourselves, but have highlighted specific aspects of that research which we hope shows what sales and trade literature can teach us.

Certainly, trade literature documents more than just business history. Here we have shown that trade literature can reveal how alliances between instrument firms and particular research scientists influenced both the development and the marketing of scientific instrumentation. Indeed, trade literature illustrates how certain scientists rose to become 'celebrities' in their own time, and how manufacturers used this celebrity to market instruments, forming a relationship that was often mutually beneficial. It is also clear that trade literature can enhance our understanding of the vibrant international culture of exhibitions that flourished after 1850. Such findings are of interest to a wide range of historians: those interested in business and commerce, in the scientific enterprise, and in broader culture as well, for after 1800 the manufacturing and marketing of scientific instruments reached a vast consuming public.[46]

---

46 One of the reasons that sales and trade literature might have been overlooked is that it can be hard to find. If anyone would like to have a look at the catalogue of sales and trade literature held in the Whipple Museum of the History of Science, which contains over 5,000 individually catalogued items (and was begun by Kemal de Soysa in his spare time), email the Museum (hps-whipple-museum@lists.cam.ac.uk) with the subject heading 'Whipple trade literature catalogue please,' and it will be send to you as an Excel spreadsheet.

CHAPTER 3

The Gentle Art of Persuasion: Advertising Instruments during Britain's Industrial Revolution

A. D. Morrison-Low

A few years ago, David Bryden examined some surviving early instruction and advertising broadsheets associated with scientific instruments. His focus was on the London makers operating between about 1650 and 1700—the years in which, as he wrote, "workshops began to change from supplying an essentially bespoke trade, to manufacturing for over-the-counter retail sales to a nobility and gentry fast recognizing that an educated man needed to be familiar with science".[1] This paper will look at another period of intensive change in the markets for instruments, that of the Industrial Revolution in the United Kingdom, usually taken to date between about 1760 and about 1830. What do the surviving ephemera from this period tell us about the makers and their targeted markets?

As Bryden observed, "by their very nature, [this] sort of printed material [...] was intended to be short lived and to serve an immediate transitory need. Having served its purpose, it would be disposed of".[2] There are, naturally, some occasional survivors, but as his papers demonstrate, these ephemera are difficult to locate being neither fish nor fowl: neither instrument nor book. This means that, by and large, they have escaped the attention of cataloguers in both libraries and museums. Bryden expressed a pious hope that with digitisation and better cataloguing, such items could be the subject of a finding aid of some sort. Certainly, with the growth of the web, there are now a growing number of finding aids for trade catalogues.[3] Instruction leaflets, however,

1 D. J. Bryden, "More Printed Ephemera of London Instrument Makers: Instructions and Advertising Broadsheets Part 1: Instructions for the Universal Equinoctial Dial, c. 1650–c. 1700", *Bulletin of the Scientific Instrument Society* 79 (2003), pp. 18–22: 18. This reworks some of the material discussed in D. J. Bryden, "Evidence from Advertising for Mathematical Instrument Making in London 1556–1714", *Annals of Science* 49 (1992), pp. 301–336.

2 D. J. Bryden, "More Printed Ephemera of London Instrument Makers: Instructions and Advertising Broadsheets Part 3: Instructions for using microscopes, 1706–c. 1730", *Bulletin of the Scientific Instrument Society* 81 (2004), pp. 10–17: 15.

3 For example, those bookmarked by the SIC: http://iuhps.org/references/onlinecatalogues.shtml (accessed 22 February 2016); besides the listings produced as publications by

 | DOI 10.1163/9789004324930_004

remain ephemera of a more hit-and-miss variety, sometimes still being found with their original instrument.

By the time that Britain was becoming industrialised, the market for scientific instruments was changing again from one aimed principally at the nobility and gentry—the members of the wealthiest classes with the most disposable income—to a more diversified and broader range of customers. These extended from the State, and its servants, through the scientist (then known as a 'natural philosopher'), to the domestic user, lecture-demonstration material, the practical tools of the navigator or surveyor, and included the now much smaller market share of the dilettante gentleman, who was usually—but not exclusively—based in the metropolis.[4]

During Britain's Industrial Revolution, the advertising of scientific instruments increased with the volume and variety of new items being produced. There was a growing population, and this in English urban areas was accompanied by a spread of literacy, together with a local press. In London, as the centre of markets and fashion, this effect could be seen by the increasing number of newspapers, which then had to be filled with news and advertisements to attract an inconstant readership.[5]

Only one large London instrument retailer appears to have survived for this entire period—the brothers William and Samuel Jones of Holborn. They had learned much from their forerunners in the trade, and alongside the expected trade catalogues, produced much more ephemeral material—leaflets, trade cards, even encyclopaedia articles. London had become the pre-eminent centre for instrument supply in the Western world by the mid-eighteenth century, but not all instruments sold in London were made there. Although the clues are slender, the evidence appears to indicate that there was a well-organised system of subcontracting, in which specialist makers supplied parts or even entire instruments to retailers—such as W. and S. Jones—who had the

---

Audrey B. Davis and Mark S. Dreyfuss, *The Finest Instruments Ever Made: A Bibliography of Medical, Dental, Optical and Pharmaceutical Company Trade Literature 1700–1939*, Arlington (Mass.), 1986, and Robert G. W. Anderson, John Burnett and Brian Gee, *Handlist of Scientific Instrument-Makers' Trade Catalogues 1600–1914*, Edinburgh, 1990.

4 Markets for scientific instrument at this period are discussed by A. D. Morrison-Low, *Making Scientific Instruments in the Industrial Revolution*, Aldershot, 2007, pp. 249–283.

5 See Brian Gee, Anita McConnell and A. D. Morrison-Low (eds.), *Francis Watkins and the Dollond Telescope Patent Controversy*, Farnham, 2014, where much of the evidence for an early eighteenth-century instrument maker's business is supported by material from contemporary newspapers.

FIGURE 3.1 *Detail of the brass rings of the Bohnenberger gyroscope, retailed by W. & S. Jones, London, c. 1835, NMS T.1990.26.*

instruments engraved with their name and place of point-of-sale.[6] There is a previously published example of a Bohnenberger gyroscope constructed for an American collector in 1831 that demonstrates this end-of-production name engraving: it is prominently inscribed on the brass rings with the retailers' particulars, and it was only by inserting an endoscope into the carved wooden globe that the name of the maker, William Stiles, was found hidden inside. He had written: "Made by W. Stiles London / June 1831 / in the Reign of King William the 4th / Commenc[in]g the glorious era of Reform" (*Pl. 4*; *Figs. 3.1* and *3.2*). Stiles worked in London, sometimes for himself—instruments are known with his signature—and possibly when times were difficult, for others who would pay him, as a jobbing contractor.[7] There is, however, absolutely no reason to doubt that without the increasingly efficient lines of communication

6 W. & S. Jones trade catalogues have been examined by Peter de Clercq, "Second-hand instruments in a W. & S. Jones Catalogue of 1795", *Bulletin of the Scientific Instrument Society* 85 (2005): pp. 38–39; and Morrison-Low, op. cit. (n. 4), pp. 164–174.

7 A. D. C. Simpson, "A Sub-contractor of W. & S. Jones Identified", *Bulletin of the Scientific Instrument Society* 39 (1993): pp. 23–27.

FIGURE 3.2 *Inside the gyroscope: detail of part of the signature of the maker, "W. Stiles, 1831", NMS T.1990.26.*

into London from the provinces, cheap instruments made in bulk could not be supplied to wholesalers from outside the city. With the growth of a progressively efficient transport network, raw materials as well as finished goods could be moved around the country, rather than taken to the nearest port and shipped to London.

For instance, a number of Sheffield instrument firms (such as J. P. Cutts) advertised themselves as 'wholesalers'—on trade cards and handbills, examples of the latter showing an engraving of their factory and a list of their products, and local directories confirming that the businesses survived for much of the period under discussion—yet vanishingly few early items have been recorded with a Sheffield or Birmingham address engraved on them (*Fig. 3.3*).

FIGURE 3.3 *Sheffield-made telescopes, signed with London 'addresses,' c. 1820, NMS T.1993.119 and 120.*

Nevertheless, items with provincial names survive with a much more fashionable London address. A slightly later commentator, the London journalist, Henry Mayhew, remarked in his 1851 survey of the *London Labour and the London Poor*, that:

> An experienced tradesman said to me: "All these low-price metal things, fancy goods and all, which you see about, are made in Birmingham; in nineteen cases out of twenty at the least. They may be marked London, or Sheffield, or Paris, or any place—you can have them marked North Pole if you will—but they're genuine Birmingham. The carriage is lower from Birmingham than from Sheffield—that's one thing".[8]

It is also more likely that customers would buy items off-the-peg in London than in a provincial centre. Wealthy patrons spent part of the fashionable year in the capital, pursuing business, legal, governmental or court activities before retiring to the country when the calendar demanded. Part of their metropolitan

8 Henry Mayhew, *London Labour and the London Poor*, London, 1861; reprinted New York, 1968, 4 vols.; v. 1, p. 333.

life was spent indulging in new consumer habits: and the growth of shopping was a part of this.[9]

Another example of more or less successful making and marketing from the English provinces to a wide, and new, domestic market is that of the kaleidoscope, invented in 1817—or so he said—by the Scottish scientist, David Brewster. Although it was supposedly protected by a patent, this process was mishandled.[10] As recounted in one provincial newspaper:

> A new optical instrument, called a Kaleidoscope, said to be the discovery and invention of Dr Brewster, of Edinburgh, is at present *all the rage* in the Metropolis—with the aid chiefly of two longitudinal darkened pieces of glass, placed at a certain angle inside a circular tube, and some common bits of coloured glass at the end—optical illusions are produced by reflection of the most beautiful colours, shape, and symmetry, and in form ever varying—such is the amusement with this simple instrument, that the Doctor's Patent is eluded and tinmen and glass-cutters are in one universal requisition—the invention is likely to be of great utility to calico and floor-cloth printers, by presenting them figures for patterns in variety of form and beauty unparalleled.[11]

This fashionable success led to a variety of contradictory advertisements in local newspapers around the country in an attempt to capture new customers while the craze for the device still raged. The Bristol press, for instance, carried an advertisement in April 1818 from a number of businesses not otherwise known to be connected with the instrument trade:

> The *Genuine Kaleidoscope* to be had only *at the Manufactory*, No. 25 Broadmead; P. Rose, Broadmead; Mintorn, College Green; Barry & Son, and Rees, High-street; Cookworthy, Corn-street; Prosser and Richardson,

---

9 See generally, Neil McKendrick, John Brewer and J. H. Plumb, *The Birth of a Consumer Society: the Commercialization of Eighteenth Century England*, London, 1982; and more particularly, Jim Bennett, "Shopping for Instruments in Paris and London", in Pamela H. Smith and Paula Findlen (eds.), *Merchants and Marvels: Commerce, Science and Art in Early Modern Europe*, New York, 2002, pp. 370–395. For the rise of shopping, see Dorothy Davis, *A History of Shopping*, London, 1966.

10 A. D. Morrison-Low, "Brewster and Scientific Instruments", in A. D. Morrison-Low and J. R. R. Christie (eds.), *'Martyr of Science': Sir David Brewster 1781–1868*, Edinburgh, 1984, pp. 58–65.

11 *Jackson's Oxford Journal*, 25 April 1818.

> Clare-street; Frost, Broad-street; Hillyard & Morgan, and Huntley, John-street, and Wood, Castle-street—*All others are spurious*.[12]

In July, a member of the Bristol instrument trade, Charles Beilby, used the local press to state that where the "Patent Kaleidoscope" was concerned, "he has extended into an agreement with Dr Brewster the Inventor and Patentee of the above very amusing Instrument, for the Manufacture of them, and that he is the only Person in the West of England who is authorised to make them [...]".[13] In the same newspaper, a Mr J. W. Hall, Broadmead, announced that he "has this day received a supply of the PATENT KALEIDOSCOPES, with the *new Improvement* as found out by the *French*".[14] By early September, despite Brewster's attempts to regulate distribution and make some profit from his design, Edward Bird and Son, Bristol watchmakers and jewellers, advertised that they had been appointed by Philip Carpenter, the Birmingham "Optician and Sole Maker of the Kaleidoscope, during the term of the patent, his Agent for the sale of these Instruments in this City; they may be had at their Shop, at various prices, from 10s to 6 guineas, accompanied with books of description".[15] Infuriated, Beilby retaliated by advertising his own instruments in the Birmingham press later that month.[16]

By the time Brewster regained some power over the situation by appointing 'approved' makers and retailers in about mid-1818, the vogue for his device was tailing off: he estimated that some two hundred thousand instruments were sold in London and Paris during the three-month peak of its popularity.[17]

12 *Bristol Journal*, 25 April 1818.

13 *Bristol Journal*, 4 July 1818.

14 Ibidem.

15 *Bristol Journal*, 5 September 1818. Examples of Carpenter's kaleidoscope, with a London address, are in the Science Museum, London, inv. nos. 1918-112 and 1990-5036/4606; Museum of the History of Science, Oxford, inv. no. 13300; and National Museums Scotland inv. no. IL.2003.44.10.42.

16 *Aris's Birmingham Gazette*, 28 September 1818. A kaleidoscope signed by Beilby is held by the Museum of the History of Science, Oxford, inv. no. 76530.

17 See [David Brewster], 'Kaleidoscope' in David Brewster (ed.), *The Edinburgh Encyclopaedia*, Edinburgh, 1808–1830, 30 vols.: v. 12, pp. 410–416 and pp. *409–*412; p.*412 gives a list of the approved sellers: "The patent kaleidoscopes are now made in London under the sanction of the Patentee, by Messrs P. and G. Dollond, W. and S. Jones, Mr R. B. Bate, Messrs Thomas Harris and Son, Messrs W. and T. Gilbert, Mr Bancks, Mr Berge, Mr Thomas Jones, Mr Blunt, Mr Schmalcalder, Messrs Watkins and Hill, and Mr Smith. In Birmingham by Mr Philip Carpenter; in Bristol by Mr Beilby; and in Edinburgh by Mr John Ruthven." The date of this has been assessed as '1818', by Charles Davies Sherborn, "Brewster's Edinburgh

Some telescopic kaleidoscopes were produced in London by jobbing instrument makers to the trade. Advertisements have been noted by "R. Huntley, No. 52½ High Holborn, Working Optician and Manufacturer of the much admired Instruments called the Kaleidoscope", and "Wrench and Co., 54 Red Lion-Street, Holborn, Kaleidoscope Manufacturers", as well as by William Harris, erstwhile partner with Brewster in a patent telescope project.[18] Although examples of the kaleidoscope signed by Harris are known, none with a Huntley or Wrench and Co. signature appears to survive, helping to confirm the idea that these businesses had no shop, but instead sold into the trade.[19] The majority of the more common telescopic instruments appear to have been manufactured in the English Midlands by wholesalers such as Philip Carpenter of Birmingham, a firm known to have been supplying telescopes to large London firms, such as the pre-eminent firm of Dollond. In Birmingham, "Mr [Philip] Carpenter soon established a large trade, and supplied even Dollond himself with large numbers of telescopes bearing his famous name".[20] However, Carpenter was able to provide only a part of the unexpectedly substantial overnight demand for kaleidoscopes, as Brewster wrote to his wife from Sheffield about his visit to

> Cam & Cutt, who have undertaken to manufacture the kaleidoscope for Mr Ruthven [of Edinburgh]. They have agreed to make and sell the instruments under my patent on the same terms as Mr Carpenter,

Encyclopaedia. Issued in 18 vols from 18– to 1830", *Journal of the Society for the Bibliography of Natural History* 1 (1936–1943), p. 112. The same text also appears in the June 1818 number of an Edinburgh journal: Anonym [but probably by D. Brewster], "History of Dr Brewster's Kaleidoscope ...", *Blackwood's Edinburgh* Magazine 3 (1818), pp. 331–337: 337. A later version of this is given, with the inclusion of "Messrs Spencer, Browning & Rust, Wapping, [...] and at Liverpool, by Messrs Egerton Smith, & Co.", in D. Brewster, *A Treatise on the Kaleidoscope*, Edinburgh, 1819, p. 167.

18 *Morning Post*, 21 May 1818; *Morning Chronicle*, 23 May 1818; *Morning Chronicle*, 17 August, 1818.

19 Kaleidoscopes by William Harris are held at the Science Museum, London, inv. nos. 1911-271 and 1931–668; and at the Museum of the History of Science, Oxford, inv. no. 74057. All of these have a folding octavo instruction handbill: "METHOD OF USING / *THE ADMIRED* / KALEIDOSCOPE, [...] Manufactured and Sold / BY / WILLIAM HARRIS, & Co."

20 [Robert Field], "Optical and Mathematical Instruments", in Samuel Timmins (ed.), *The Resources, Products, and Industrial History of Birmingham and the Midland Hardware District*..., London, 1866, p. 534.

FIGURE 3.4 *Kaleidoscope, signed by I. Ruthven, Edinburgh, manufactured by Cam & Cutts, Sheffield, 1820,* NMS *T.1985.26.*
© NATIONAL MUSEUMS SCOTLAND.

> provided I get his permission to allow them to be employed. This I must do, as he cannot possibly supply the demand.[21]

Examinations of dismantled examples show that the ends of the kaleidoscope, which are stamped pieces of brass, could easily have been produced by, for instance, a button manufacturer: "the tinmen and the glaziers began to manufacture the detached parts of it, in order to evade the patent," as Brewster complained.[22] The partnership between James Cam and John Cutt or Cutts lasted only one year, that of 1818, and was probably brought about solely to deal with this single lucrative contract (*Fig. 3.4*). The example of the kaleidoscope demonstrates an instance of an instrument—really no more than a toy—that rapidly captured the public imagination, but did not require enormous skill to construct. The rapid appearance of a new domestic market, one in which aspirational people used instruments in the privacy of their drawing rooms, was a direct result in the rise of a consumer society, one of the by-products of the Industrial Revolution. Understandably, David Brewster had failed to anticipate this novel market demand, and despite his precaution of a patent there was not yet the means to harness market forces in favour of the inventor.

21 Margaret M. Gordon, *The Home Life of Sir David Brewster*, Edinburgh, 1869, p. 96. A kaleidoscope by Ruthven is in the Science Museum collection, inv. no. A680739, and inv. no. NMS T.1985.20.

22 [Brewster], op. cit. (n. 16), p. *410.

The press proved to be a magnificent method of persuading the newly-literate classes that fashionable things they could afford were available; but it was not the only way to influence people to acquire instruments new or second-hand. Booklets and broadsheets, made to go with instruments, or to be bound into the back of books, flourished at this period, and here the microscope is probably the most prominent example. The microscope was a device with optics that remained relatively unimproved until the 1830s, although the instrument itself dated from the late seventeenth century. Thus, in order to make it fashionably appealing, makers applied as much ingenuity as they could to its mechanical structure during this period. Rather than looking at the more expensive end of what was available, an example from the lower end of the market offers a neat case in point: this is what became known as the Cary-Gould microscope, famous for being taken into the field and used by botanists, clergymen, amateurs to a man (or woman). It proved to be one of the most successful of the non-achromatic nineteenth century designs.

William Cary was apprenticed to Jesse Ramsden, the pre-eminent scientific instrument maker of his generation, probably between about 1773 and 1780. By 1789 Cary had set up in his own business on the Strand, and successfully made and sold astronomical and mathematical instruments. Cary died in 1825, just as one of his workmen, Charles Gould, designed a pocket microscope for exactly that emerging market of amateur natural historians.[23] It was initially described in a brief undated leaflet of only three sides—the earliest so far located dates from about 1825.[24] Cary's firm was continued in William Cary's name by his nephews George and John, so that, as Gerard Turner explains, "this type of instrument has become known as 'a Cary-type microscope' although 'Gould-type' would be fairer".[25] Gould published a booklet in 1827 entitled *The Companion to the Microscope*, and by 1848 it had run to 15 editions. The engraved frontispiece shows the parts of the instrument of Gould's "Improved Pocket Compound Microscope". Gould also, through the publisher William Cary, produced hand-bills to advertise both the book and the instrument.

National Museums Scotland has recently acquired a microscope of this type, with what appears to be an almost complete paper record of its design,

---

23 Brian Stevenson, "The Cary-Gould-Porter optical businesses", see http://www.microscopy-uk.org.uk/mag/artjan14/bs-Cary_Gould_Porter.pdf (accessed 22 February 2016).

24 R. H. Nuttall, *Microscopes from the Frank Collection 1800–1860*, Jersey, 1979, p. 60: this example in the British Library, London.

25 Gerald L'E. Turner, *The Great Age of the Microscope: The Collection of the Royal Microscopical Society through 150 Years*, Bristol and New York, 1989, p. 75.

manufacture, and initial sale—a good example, indeed, of how instruments change hands (at least at the start of their existence; *Pl.* 5, *Figs.* 3.5 and 3.6). However, this seemingly complete documentation may not be as straightforward as it appears. First, there is the instrument itself, which is what is described in the second part of the accompanying surviving leaflet, the "Description" (the second piece of evidence). This categorically states that one "A. McKenzie" "manufactured and sold" this item. Who was "A. McKenzie"? Did he really *make* anything at all? Turning to the trusty bible (by Gloria Clifton), she states that Alexander MacKenzie (or McKenzie) was working between 1818 and 1830, and was at 15 Cheapside between 1816 and 1822. The firm is known to have sold a magnetic compass and another microscope; hardly enough to make the case for the existence of a large manufactory producing brass and glass parts, putting them together and providing individual polished wooden cases for each one.[26] Third, there is a brass plate inset into the lid of the box, for—presumably—the first proud owner, one Robert Green. Perhaps Mr Green was given this as a prize, maybe as a present; perhaps it was just an object that he cherished. However, although we may hesitate to believe that Mr McKenzie 'made' the instrument, he undoubtedly sold it.

In his *Great Age of the Microscope*, Gerard Turner comments that the signatures of three London addressed Cary-Gould microscopes appear to have been signed by the same hand. Two were sold by Cary, the third by R. B. Bate: but, as Turner says, "it is likely that one manufactory supplied the entire English trade in these pocket microscopes".[27] So where might that manufactory have been located?

By 1830 it appears that instruments were being produced in the new industrial centres of the English Midlands, particularly in Sheffield (long-term home to the cutlery trades) and Birmingham (where small brass items were made), and moved in large numbers, first by packhorse, then by canal and subsequently by railway, into the great London market-place. There they could be safely inscribed by any retailer's name, without the customer knowing where the item had originated. After the Great Exhibition of 1851, however, those selfsame provincial manufacturers—for example I. P. Cutts of Sheffield (the same maker as James Cutts, mentioned earlier), who had London premises at Crown Court, Fleet Street from at least 1836—were happier to put their own name and

26 Gloria Clifton, *Directory of British Scientific Instrument Makers 1550–1850*, London, 1995, pp. 175 and 186.

27 Turner, op. cit. (n. 21), p. 76. Turner does not hazard a guess as to where this manufactory might be.

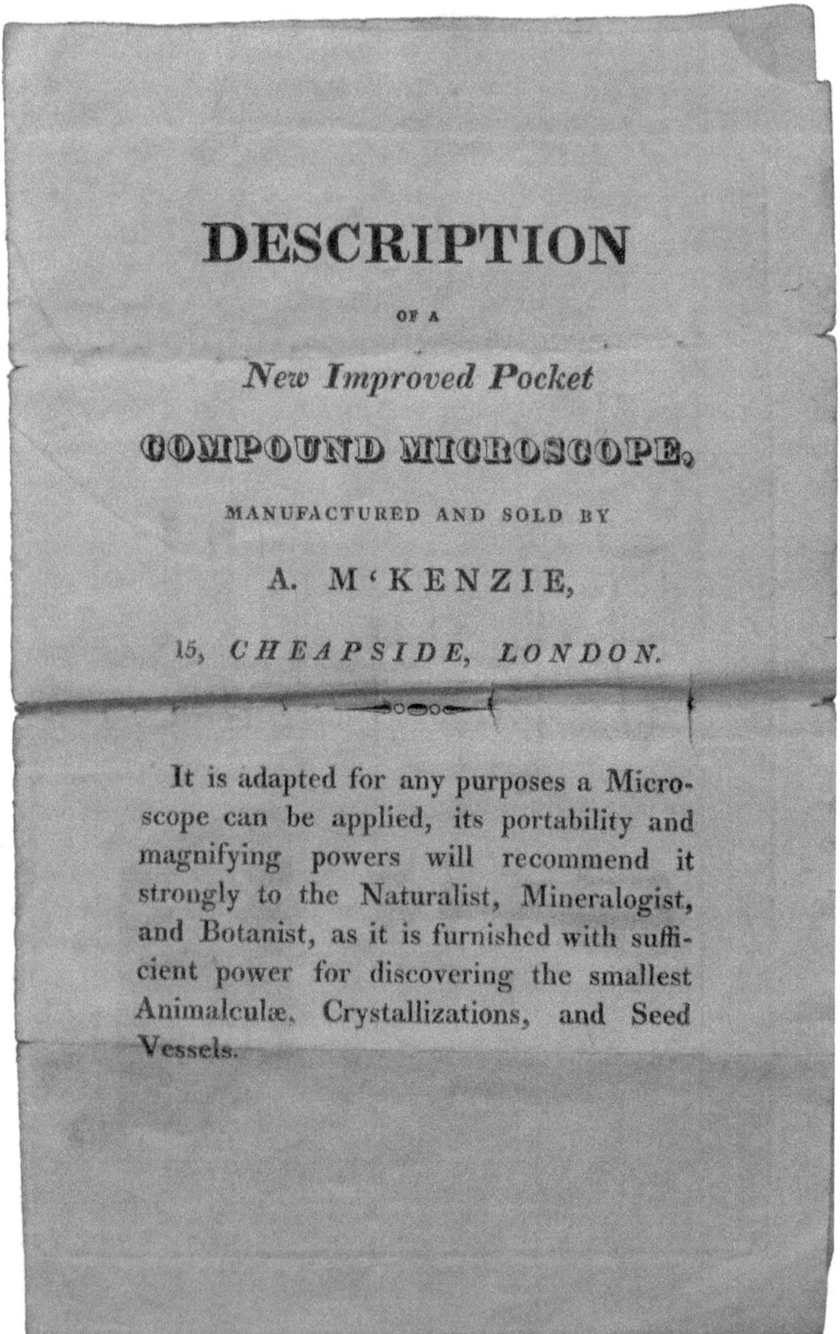

DESCRIPTION

OF A

New Improved Pocket

COMPOUND MICROSCOPE,

MANUFACTURED AND SOLD BY

A. M'KENZIE,

15, *CHEAPSIDE, LONDON.*

It is adapted for any purposes a Microscope can be applied, its portability and magnifying powers will recommend it strongly to the Naturalist, Mineralogist, and Botanist, as it is furnished with sufficient power for discovering the smallest Animalculæ, Crystallizations, and Seed Vessels.

FIGURE 3.5 *Leaflet with Cary-Gould microscope, written by Alexander McKenzie,* NMS *T.2008.170.* © NATIONAL MUSEUMS SCOTLAND.

FIGURE 3.6 *Inset brass plate with owner's name, "Robert Green" on Cary Gould microscope, NMS T.2008.170.*

provincial address on their instruments and produce leaflets to instruct their customers. Another Sheffield-based firm, with a Liverpool branch, Chadburn, similarly advertised their form of Cary-Gould microscope design from 1845, and began to sign them. On the way, these wholesalers had persuaded their customers that provincial instruments were not poor quality; in fact, in the case of Chadburns, that the cheapness of their scientific instruments was a positive attribute.[28]

By the time most historic scientific instruments reach museums (usually the end of changing hands, as most British museums are unable to de-accession collections), their origins and the journey from those beginnings, have become anonymous. Unlike paintings, where provenance is of paramount importance in proving that an item is what it purports to be, instruments—especially those from the lower end of the market, such as those just discussed—have usually lost their history by passing through a saleroom at some point in their history. However, the instruments discussed in this paper have possibly had a kinder fate. Some of the Bohnenberger gyroscope's history has been uncovered, as have the beginnings of the Ruthven kaleidoscope and the provincially-made, metropolitan-sold telescopes. The Alexander McKenzie microscope

28 Chadburn Brothers of Sheffield and Liverpool made use of the phrase 'Good and Cheap Instruments' in their advertisements after the Great Exhibition in 1851, and again after the 1862 International Exhibition in London.

was acquired from an antiquarian bookseller (he was adamant that the information in the pamphlet was gospel truth), and the other Cary-Gould microscopes came from the Frank Collection. Purchased by the collector Arthur Frank in the early 1970s in the major London salerooms, these came to National Museums Scotland in 1979. Mr Frank paid many times their original price new, but considerably less than would be paid today.

CHAPTER 4

Some Considerations about the Prices of Physics Instruments in the Nineteenth Century

*Paolo Brenni**

Introduction

In the past few decades, many articles and books have been written about nineteenth century scientific instrument makers, their products, and the instrument trade.[1] It is well known that the most important makers of this period were found in England, France, and Germany, but their influence on and penetration into the global market of instruments varied. Between the first decades of the eighteenth century and the beginning of the nineteenth, only English makers could provide the best astronomical, optical, surveying, and navigational instruments. French precision industry flourished after the Revolution. With great skill and inventiveness, Parisian makers flooded the international market with their products from the 1820s until the end of the century. Starting in the 1870s, the German precision industry boomed, profiting from and contributing to the formidable scientific and industrial development of the new Reich. It began to be a fierce competitor for France and the United Kingdom by producing an increasingly important volume of excellent instruments. Since the end of the eighteenth century, the United States imported very large amounts of instruments from Europe, and only after World War I did their domestic industry seriously compete with that of the Old Continent.

* I would like to thank several colleagues who, with their suggestions posted to the mailing list *Rete*, greatly helped me in my research. I am particularly grateful to Richard Kremer, who was always ready to answer my numerous and pressing questions and made several valuable comments concerning this article.

1 For a general introduction on the history of the instrument-making industry see for example: Anita McConnell, "Instruments and Instrument-Makers, 1700–1850", in Jed Z. Buchwald, Robert Fox (eds.), *The Oxford Handbook of History of Physics*, Oxford, 2013: pp. 326–357 and Paolo Brenni, "From workshop to factory: The evolution of the Instrument-Making Industry, 1850–1930", in Ibid.: pp. 584–650. See also: Peter R. de Clercq (ed.), *Nineteenth-century Scientific Instruments and Their Makers: Papers Presented at the Fourth Scientific Instrument Symposium, Amsterdam 23–26 October 1984*, Amsterdam, 1985.

 | DOI 10.1163/9789004324930_005

Several factors influenced the popularity and the diffusion of instruments coming from different countries: First of all was their quality and their design. But how much did prices influence the worldwide diffusion of instruments made in England, France, Germany, or, later, the United States?[2] Research has been made on some of the most important European and American workshops and their organisation. However, very little has been written concerning the cost of instruments. How did costs evolved over time? Were they stable? Did they differ, for the same type of instrument, from one maker to another? Were their prices the same in different European countries and in the United States? Were preferences of one instrument maker over another due to his prestige and renown or was it also dictated by the cost of his products? For American customers, was it more convenient to acquire instruments directly from Europe or did they prefer local makers and retailers? These comparisons, together with an analysis of price changes, will be useful to cast some new light on several aspects of the instrument trade during the nineteenth and the early twentieth centuries.

It should be noted that it is much easier to compare instrument prices in this period than in previous ones. In fact, during the nineteenth century, especially after about 1850, many kinds of apparatus acquired a mature and definitive design, which was reproduced over and over with the same materials and without major modifications. This is particularly true for physics instruments, which equipped the increasing number of schools, universities, and various laboratories that were created or completely refurbished in the nineteenth century. These instruments were described, illustrated, and also implicitly advertised by dozens of physics treatises, which aided in their diffusion (and the practices associated with them) both in Europe and in the United States. What can be seen as a kind of standardisation of physics apparatus led most instrument makers to manufacture or offer nearly identical instruments. Furthermore, many makers and retailers were selling instruments made by others under their own brand signatures. This practice was very common at least until the beginning of the twentieth century.[3] Because a price

2 In this article I only consider England, France and Germany because, during the nineteenth century, they were by far the most important producers and exporters of scientific instruments. Until the beginning of the twentieth century the United States were essentially importers of instruments, but it is interesting to analyse the prices of European instrument sold by American instrument makers and retailers in the USA.

3 P. Brenni, "Who made what? The dilemma of signed instruments", in Vasilii P. Borisov et al., *XVIII International Scientific Instrument Symposium Moscow-St. Petersburg, Proceedings, Russia, 20–25 September 1999*, Moscow, 2002: pp. 50–66.

comparison makes sense only when it considers identical or very similar objects, I have chosen for my analysis various instruments that were common in physics cabinets and laboratories of the time, such as electrostatic machines, vacuum pumps, induction coils, and several other types of research and teaching apparatus. Although a comparison is quite easy for physics instruments, because almost every maker produced the same, identical artefacts, such an analysis is more difficult, and often quite impossible, for astronomical, navigational, or surveying instruments. Setting aside sextants, which differed little from one maker to another, instruments such as theodolites often maintained characteristics of tacitly-established national styles, in spite of the fact that they all performed exactly the same operations and measurements. For example, British theodolites often had four levelling screws for feet, instead of the common three screws used in France, or they had a different type of divided circle or surface finish. For these reasons, the same class of apparatus could have different prices because of different designs, construction details, and materials. The same can be said about microscopes or telescopes, often sold with different accessories and optics by different makers. All these details certainly influenced the costs of manufacturing and therefore the retail prices.

The Trade Catalogues: An Indispensable Tool of Research

Instrument makers' trade catalogues are the essential tools for this research. Until a few years ago these catalogues, which were considered 'grey literature' of lesser importance, were very difficult to find. They were stored in obscure sections of libraries, hardly considered by most booksellers, and often forgotten in some dusty cupboard in the cellars or in the attics of scientific institutions. After the 1980s, the increasing interest in scientific instruments made trade catalogues valuable tools for instrument historians and collectors. Today, thanks to the Internet, a large number of catalogues can be easily found online.[4]

We must be careful, however, in studying catalogues. It is well known that several makers reproduced in their catalogues the illustrations published in

4 Hundreds of trade catalogues of instrument makers can be found in the reference webpage of the Scientific Instrument Commission: www.sic.iuhps.org/references/online catalogues.shtml (accessed 24 February 2016).

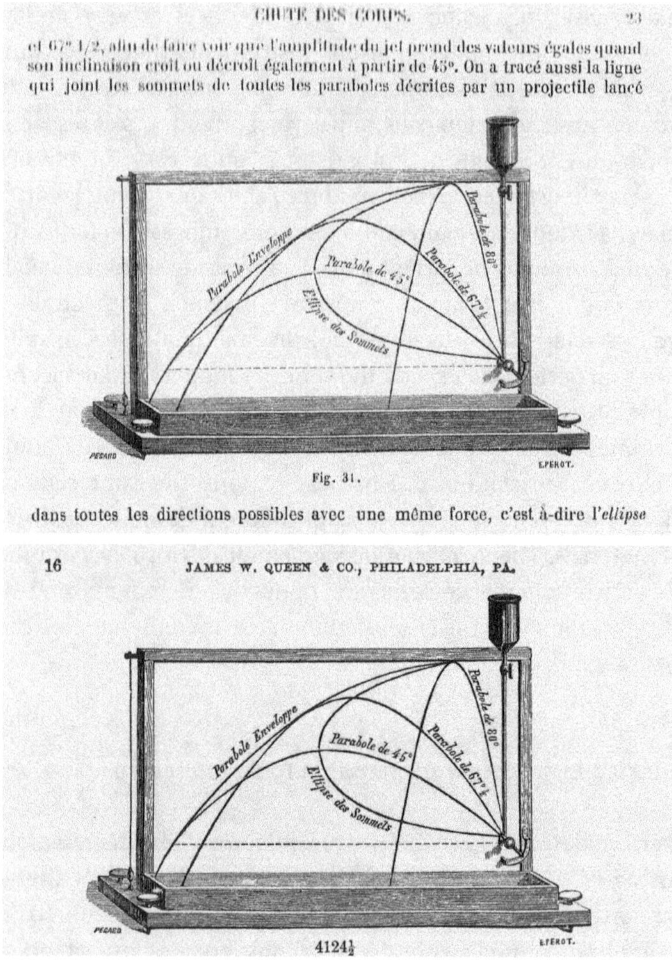

FIGURE 4.1 *The apparatus for demonstrating the parabolic trajectory of liquids from the catalogue of J. Salleron* (Notice…, *Paris, 1864, p. 23), and the same illustration reproduced twenty-four years later in the catalogue of James W. Queen and Co.* (Priced and Illustrated Catalogue…, *Philadelphia, 1888, p. 16*).

competitors' catalogues or found in scientific treatises and articles (*Fig. 4.1*).[5] This practice was common until the beginning of the twentieth century.

5 P. Brenni, "The Scientific Instrument Makers Illustrated Catalogues", in Christine Blondel, Françoise Parrot, Anthony J. Turner (eds.), *Studies in the History of Scientific Instruments, Papers presented at the 7th Symposium of the Scientific Instrument Commission, Paris 15–19 September 1987*, London, 1989, pp. 169–178.

Several factors encouraged it. Not only could a maker avoid having new and expensive printing blocks expressly made for his own catalogues, but he could also illustrate instruments that would be manufactured only by custom order or purchased from other makers and then resold. Very often the instruments sold were not exactly like the ones illustrated in the catalogues. Nevertheless, if we consider very common instruments that were technically mature and had a well-established design (such as many physics apparatus), we find that their material differences were minimal and a comparison of their prices is possible.

Generally, the catalogues printed before about 1850 were simple lists of instrument names and prices, sometime enriched with a few additional details (such as the diameters of glass plates on the electric machines; *Fig. 4.2*). When we compare these catalogues, it is necessary to select only instruments whose design and characteristics were well defined, so that a simple denomination in different lists clearly refers to a well-known, particular type of instrument—for example, a French-type friction electrostatic machine with double prime conductors, or a double-barreled vacuum pump.

In the second half of the nineteenth century most catalogues were illustrated. If we trust these illustrations, it is easier to select comparable instruments made by different firms. Furthermore, in this study, I selected the catalogues of some very important manufacturers, who almost certainly manufactured (and did not simply resell) the instruments that they advertised. When it was possible, I also took advantage of the physics collection of the Fondazione Scienza e Tecnica in Florence in order to double-check the price of various instruments. Many instruments preserved in this large collection were made by some of the most important French and German firms of the time. In many cases the prices the school actually paid for individual pieces of apparatus were recorded in registers and inventories, which still survive. When I checked these registers, I found that prices paid matched those published in maker's trade catalogues.

Currencies and Exchange Rates

When we study the prices of instruments made in different countries, it is necessary to know the exchange rates of different currencies during the considered period of time. In 1865 France, Italy, Belgium, and Switzerland formed the Latin Monetary Union, which was based on common gold and silver standards. Their national currencies became interchangeable (1 French Franc = 1 Swiss Franc = 1 Italian Lira, etc.). A few years later Spain and Greece joined the Union and were later followed by other European countries. The Union worked

(19)

cristal, avec pistolet de Volta sur le chapiteau, et un tabouret isolant. 2000

Nota. Cette machine, ainsi que les suivantes, est garnie d'armatures en taffetas qui en triplent la force.

268. Machine électrique à plateau de 36 pouces, semblable à la précédente. 900

269. Machine *idem*, de 32 pouces. 700

270. Machine *idem*, de 30 pouces. 600

271. Machine *idem*, de 24 pouces. 450

272. Machine *idem*, de 20 pouces. 350

273. Machine *idem*; de 18 pouces, à un seul conducteur 220

274. Machine *idem*, de 18 pouces, sans table. . 190

275. Machine *idem*, de 15 pouces. 140

276. Machine *idem*, de 12 pouces 100

277. Machine électrique de Nairne, à cylindre, montée sur sa table, à deux conducteurs; le cylindre a environ 12 pouces de diamètre sur 18 pouces de long. 400

278. Machine *idem*, d'environ 8 pouces sur 12, sans table. 250

279. Deux grands conducteurs en fer-blanc peint, de 6 pieds de longueur, avec les tringles de réunion pour suspendre au plafond. . . 60

280. Deux conducteurs *idem*, plus petits, d'environ 4 pieds. 40

281. Un tabouret isolant, d'environ 16 pouces. carrés. 20

282. Un tabouret *idem*, plus grand, pour placer un fauteuil 36

283. Bouteille de Leyde, garnie, pour donner la commotion, différentes grandeurs. de 2 à 5

284. Bouteille de Leyde, étincelante, d'aventurine. . . . . 2 à 5

285. Bouteille de Leyde, à araignée, de Franklin. 10

286. Bouteilles aux trois étincelles. 10

287. Appareil pour l'analyse de la bouteile de Leyde. 12

2.

FIGURE 4.2 *A page of the 1821 Pixii's catalogue listing various electrical machines of different sizes and types.*

PIXII, *NOTICE DE DIFFÉRENS* [SIC] *INSTRUMENTS . . .*, PARIS, 1821, P. 19.

relatively well until the beginning of the twentieth century and finally collapsed during World War I. Its official end came in 1927.[6]

Until World War I, a British pound corresponded to 25.20 French Francs (and the associated currencies). The exchange rate of the U.S. Dollar to the British Pound was also quite stable during the nineteenth century. During this period, £ 1 corresponded to $ 4.85. Only during the years of the American Civil War and the post-war decade did the exchange rates vary dramatically, reaching a maximum of $ 10 for £ 1. The pre-war rate was reached again in the mid-1870s. The German gold Mark was introduced in 1873, a couple of years after the foundation of the German Empire. One British Pound corresponded to 20.43 Marks.

Compared to the huge fluctuations experienced by the currencies after World War I and for the rest of the twentieth century, we certainly can affirm that, apart from some short periods often corresponding to wartime events (for example, the American Civil War and the Franco-Prussian War), exchange rates between the end of the Napoleonic era and 1914 were extremely stable. Moreover, small fluctuations on the order of a few cents in the exchange rates—perhaps important for bankers, financiers, and investors—were not important factors in instrument sales, and do not affect my general analysis. In the table below, I have summarized the average values for the exchange rates of the most important currencies.[7] These are the values I used when comparing the prices of instruments made in different countries.

TABLE 4.1 *Average relative value of different currencies: French Franc (Fr), German Mark (Mk), U.S. Dollar ($), English Pound (£)*

	Fr	Mk	$	£
Fr	1.00	0.81	0.19	0.04
Mk	1.23	1.00	0.24	0.05
$	5.20	4.20	1.00	0.17
£	25.20	20.43	5.87	1.00

6 See: Henry Parker Willis, *A History of Latin Monetary Union*, Chicago, 1901. See also: https://courses.cit.cornell.edu/wbb1/papers/Bae%20Bailey%20LMU-2003.0707.pdf and http://en.wikipedia.org/wiki/Latin_Monetary_Union (both accessed 24 February 2016).

7 See: Markus A. Denzel, *Handbook of World Exchange Rates, 1590–1914*, Farnham, 2010; Samuel T. Morgan, *U.S. Import Duties under the existing laws and decisions and digest of tariff laws*, Baltimore, 1883: exp. "Appendix", pp. 1–6; and also www.measuringworth.com and http://projects.exeter.ac.uk/RDavies/arian/llyfr.html (both accessed 24 February 2016).

The Prices of French Physics Instruments

French instruments, and particularly philosophical instruments, were highly appreciated during the nineteenth century. Not only were they largely exported to several European and American countries, but they were also taken as models to be copied by several manufacturers outside of France. To begin I will examine the variations in prices of a few instruments made by the same maker over a relatively long period of time. Obviously, such an analysis requires not only to find makers who were in business for at least several decades, but also to have access to a series of their catalogues spanning this period of time. I will take as an example three successful and long lasting Parisian firms, whose products are well suited for comparison: Pixii, Nachet and Koenig.

In 1818, Nicolas Constant Pixii (1776–1861) took over the famous workshop of the Dumotiez brothers, which had been active since about 1780. During the first half of the nineteenth century, Nicolas and his son Hippolyte, the inventor of the well-known magneto-electric generator, were the most important French makers of philosophical instruments. Not only did they equip dozens of collections in France, elsewhere in Europe, and in the United States, but they also were responsible for establishing what are now seen as the 'classical apparatus', which equipped many physics laboratories until the beginning of the twentieth century. I have chosen four typical instruments produced first by Dumotiez and then by Pixii between 1817 and 1852.[8] These instruments could be found in almost all physics collections of the time. They are a pneumatic pump, a friction electrical machine, a polyprism with stand and an hydrostatic balance. The size of the pumps was determined by the diameter of their receiver plates; that of the electric machines by the diameter of their glass disks.[9]

8 Dumotiez, *Catalogues des instruments servant à l'électricité*, Paris, [c. 1809]; Dumotiez, *Catalogue des différens* [sic] *instruments de physique, d'optique et de mathématiques*, Paris, 1817, and Pixii, *Notice de différens* [sic] *instruments de physique, d'optique et de mathématiques*, Paris, 1821; Pixii, *Catalogue des principaux instruments de physique, d'optique et de mathématiques*, Paris, 1828; Pixii père et fils, *Catalogue des principaux instruments de physique, chimie, optique et mathématiques*, Paris, 1842 and also the editions of 1849 and 1852.

9 It is the disk machine of French design with double prime conductor. In continental Europe this machine was traditionally (but erroneously) named the "Ramsden machine".

TABLE 4.2 *Four typical physics instruments with their makers and their prices in French Francs for different years*

	Dumotiez				*Pixii*			
	1809	*1817*	*1821*	*1828*	*1842*	*1845*	*1849*	*1852*
Pneumatic pump (receiving plate of 32–33 cm)		800	850	850	800	800	800	750
Electrical machine (glass disk of 80 cm)	650	600	600	650	650	650	650	650
Polyprism with stand		45	45	45	45	40	40	40
Hydrostatic balance		500	500	500	450	450	450	450

The second example concerns three instruments made in the second half of the nineteenth century by the Maison Nachet. This firm was founded by Camille Sébastien Nachet (1799–1881) in 1839, and it specialised in microscopes and apparatus for microscopy. It was very successful and survived until the late twentieth century. My examples include an inverted (chemical) microscope (*Fig. 4.3*), a camera lucida, and a dissecting microscope. They were typical products of the firm, which were offered in its trade catalogues for several decades.[10] It is difficult to compare the prices of other instruments by Nachet because they altered the design of microscopes and ancillary apparatus and improved them.

The third example concerns a few acoustical instruments made by the German-born maker Rudolph Koenig (1832–1901), working in Paris.[11] For almost fifty years Koenig was not only the most famous manufacturer of this kind of apparatus, but he also invented or improved many acoustical instruments. His experimental researches on acoustics were widely appreciated and discussed.

10 Gerard L'E. Turner (introduction by), *Maison Nachet Catalogues des Fonds de 1854 à 1910*, Paris, 1979. The volume is a reprint of 12 different catalogues of the firm Nachet.

11 Rudolph Koenig, *Catalogue des Appareils D'Acoustique*, Paris, 1865, and also the 1873 and 1889 editions of the same catalogue.

16. — **Microscope renversé pour les études de chimie** (fig. 16). — Dans cet instrument, les objectifs étant placés sous l'objet, on n'a pas à craindre que les évaporations empêchent la vision nette par suite de l'accumulation des vapeurs. — La platine

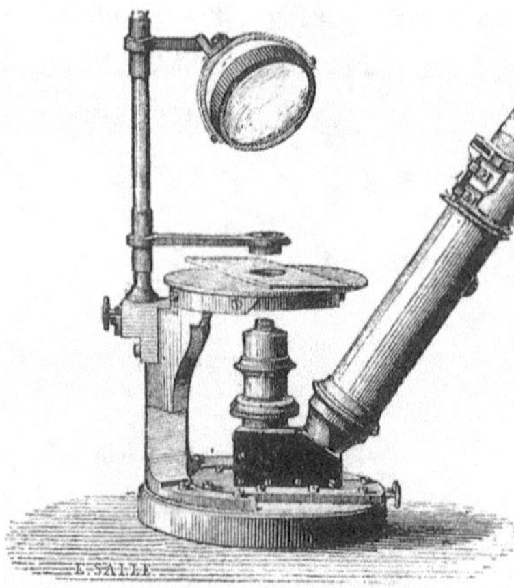

Fig. 16.

est dorée. Ce microscope, avec 4 objectifs n°s 2, 3, 5 et 6.—1 oculaire mobile. — Goniomètre pour mesurer les angles des cristaux. — Accessoires : lampe à alcool montée, sur pied articulé, lames de verre creuses, lamelles minces. Dans une boîte d'acajou. **350** fr.

FIGURE 4.3 *An inverted chemical microscope advertised by Nachet at the same price for more than half a century.*
NACHET, *CATALOGUE DESCRIPTIVE . . .*, PARIS, 1881, P. 19.

TABLE 4.3 *The prices in French Francs of three instruments made by Nachet between 1856 and 1910. The increase in the price of the camera lucida was due to an improvement of the instrument introduced in 1881*

	1856	*1862*	*1872*	*1881*	*1886*	*1892*	*1900*	*1910*
Inverted microscope	350	350	350	350	350	350	350	350
Camera lucida	25	25	25	30	30	30	30	30
Dissecting microscope				60	60	60	60	60

TABLE 4.4 *The prices in French Francs of four instruments made by Koenig between 1865 and 1889. The increases were due to the introduction of some changes and improvements in the design of the apparatus*

	1865	*1873*	*1889*
Set of 10 resonators	100	110	110
Helmholtz double siren	400	450	450
Optical comparator	120	130	140
Harmonic analyser (8 resonators)	250	300	325

From these examples it is apparent that the prices of several instruments did not change, or they varied very little (no more than 10%) over half a century. Many other cases would be very similar. This fact might seem quite extraordinary to us. First, after World War I—and even more during the second half of the twentieth century—the speed of technological progress and market pressures made it impossible for any firm to sell the same instrument without modification for thirty, forty, or fifty years. Furthermore, economic fluctuations in inflation in the twentieth century would not have allowed such price stability.

But were the prices proposed by different makers homogeneous? Did the same instruments made by different makers have different prices? I selected the following instruments: a Melloni bench for studying 'radiant heat', a double-barreled vacuum pump (diameter of the receiver plate, 32 cm), a large friction electrostatic machine (diameter of the glass disk, 80 cm), and the largest available version of Atwood's machine (*Fig. 4.4*). Again such examples were selected from the catalogues of some of the best Parisian makers, including the above mentioned Pixii,[12] Molteni,[13] Salleron,[14] Deleuil,[15] Lerebours et Secretan,[16] and

12 Pixii, op. cit. (n. 8).

13 J. Molteni et C.ie, *Catalogue et prix courant des principaux produits*, Paris, 1859.

14 Jules Salleron, *Notice sur les instruments de précision Troisième et quatrième partie Pesanteur, hydrostatique, calorique, mécanique*, Paris, 1864.

15 Deleuil, *Catalogue des Instruments de physique, de chimie, et de mathématiques*, Paris, 1863.

16 Lerebours et Secretan, *Catalogue et prix des instruments de physique, de chimie, de mathématiques, d'astronomie et de marine*, Paris, 1853.

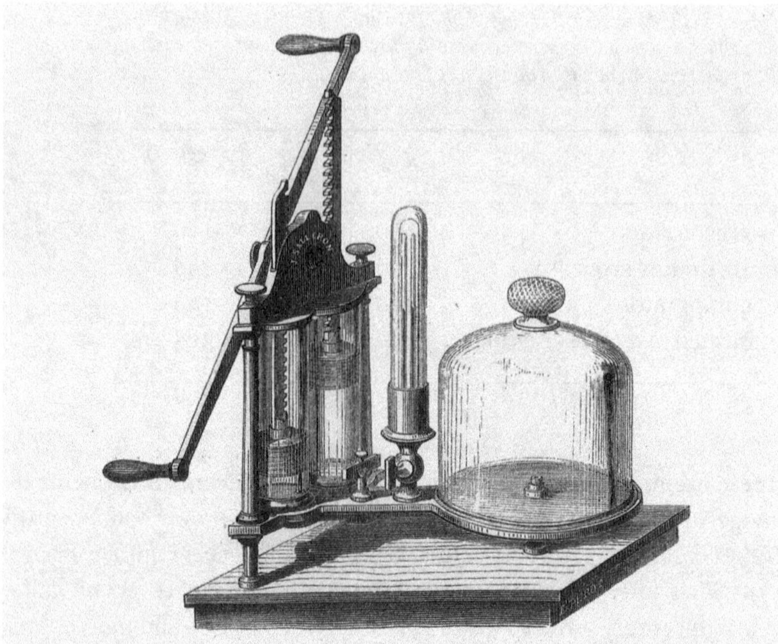

Fig. 141.

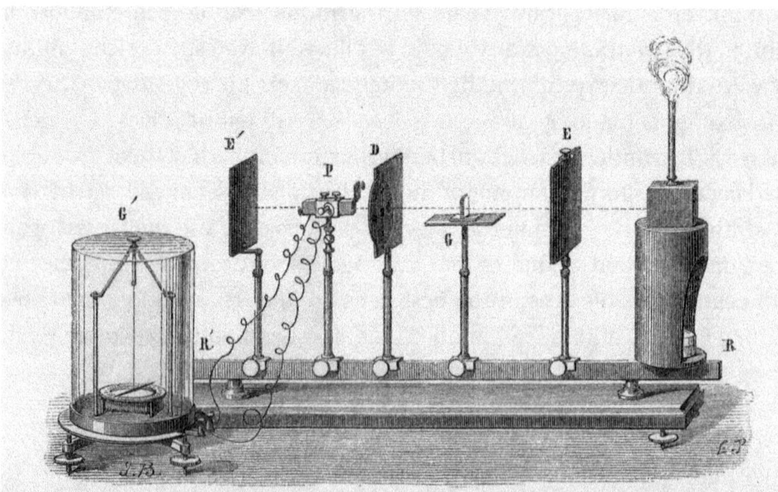

Fig. 377.

FIGURE 4.4 *A double-barreled vacuum pump and a Melloni bench. The two instruments were reproduced for several decades without modification by several instrument makers.*
SALLERON, *NOTICE* . . ., PARIS, 1864, PP. 96 AND 262.

Breton.[17] All these makers were esteemed both in France and abroad, and their instruments certainly represent the top of French production.

Table 4.5. clearly shows that the prices were quite homogeneous. These figures also show that the differences in price were in the order of 10% to 15%, and these differences were due to slight construction differences in the execution of the 'standard' instruments. Only the Atwood's machine offered by Lerebours et Secretan was really more expensive, but I suspect that it had some special features that were not present in those of other makers.

The makers mentioned were among the ones who were able to export a large percentage of their production. They were responsible for the fame of the French, precision-instrument industry during the nineteenth century all around the world. Very often a university or a school willing to equip or update its laboratories and cabinets did not purchase just a single instrument but rather ordered a whole set of them. In such cases the financial differences could disappear; the price of a slightly more expensive instrument would offset the lower price of another apparatus. Considering the fact that the differences in price were small, and, as far as we can see, no good maker was systematically

TABLE 4.5 *The prices in French Francs of four quite expensive instruments from the trade catalogues of different Parisian makers. The missing prices indicate that the corresponding maker was not selling the instrument or proposed a quite different version of it*

	Pixii	*Molteni*	*Salleron*	*Deleuil*	*Lerebours et Secretan*	*Breton*
	1852	*1859*	*1864*	*1865*	*1853*	*1867*
Melloni bench with accessories		675	650	700	650	650
Vacuum pump (plate 32 cm)	750	850	700	700	900	650
Electric disk machine (80cm)	550	575		600	600	600
Large Atwood's machine	800	700	800	700	1000	700

17 Breton, *Catalogue et prix des instruments de physique, de chimie, d'optique, de mathématique et d'astronomie*, Paris, 1867.

more expensive than the others, the choice of a supplier was hardly dictated by the prices of his instruments. Considering only high quality instruments, we must admit that other reasons prompted this choice. Good personal relationships and the confidence of the customer in the maker,[18] prestige and renown of the latter, immediate availability of the desired instruments, punctuality in delivering the ordered apparatus, these were probably some of the most important factors that made a client prefer a certain supplier.

Prices of Physics Instruments in Europe

I will try now to compare the prices of some instruments made by a few well-known English, French, and German makers between the end of the nineteenth century and the beginning of the twentieth. In fact in this period several of the most widely-diffused instruments had reached their maturity and were produced with almost no modifications by several important firms. In spite of this uniformity, some types of instruments took on special characteristics favoured in particular countries. Moreover, some instruments manufactured and widely used in certain countries, were not commonly made and used in others. For example Wiedemann's galvanometers were routinely employed in Germany, but not (or much less) in France. By contrast, the Pellat's absolute electrodynamometer was promoted and sold in France (by Carpentier and Ducretet), but not in Germany. Other more sophisticated instruments, such as certain types of polarimeters or spectrometers, were made only by a few makers, each having their own special designs and characteristics. Therefore, a comparison between such apparatus made by different manufacturers is practically impossible.

For *Table 4.6*, I have chosen various instruments, which were quite common in laboratories around 1900 and thus promoted by different makers. The makers were the German firms of Max Kohl AG of Chemnitz,[19] Leybold's Nachfolger of Köln,[20] and F. Ernecke of Berlin;[21] the French firm of Ducretet;[22]

18 It is well known that many scientists had their preferred makers. But also schools and universities tended to be faithful to one or a few makers for their regular acquisitions.

19 Max Kohl AG, *Preisliste Nr. 50 Band II und III Physikalische Apparate*, Chemnitz, [c. 1911–1912].

20 E. Leybold's Nachfolger, *Eirichtngen und Apparate für den Physikalischen Unterricht sowie für Übungen im Praktikum*, Cöln a. Rhein, [1904].

21 Ferdinand Ernecke, *Preisliste N.18 Physikalische Apparate*, Berlin, [c. 1905].

22 Eugène Ducretet, *Catalogue raisonné des instruments de précision, Première et deuxième parties Physique générale*, Paris, 1905, and E. Ducretet et L. Lejeune, *Catalogue des instruments de précision, Troisième partie, Electricité*, Paris, 1892.

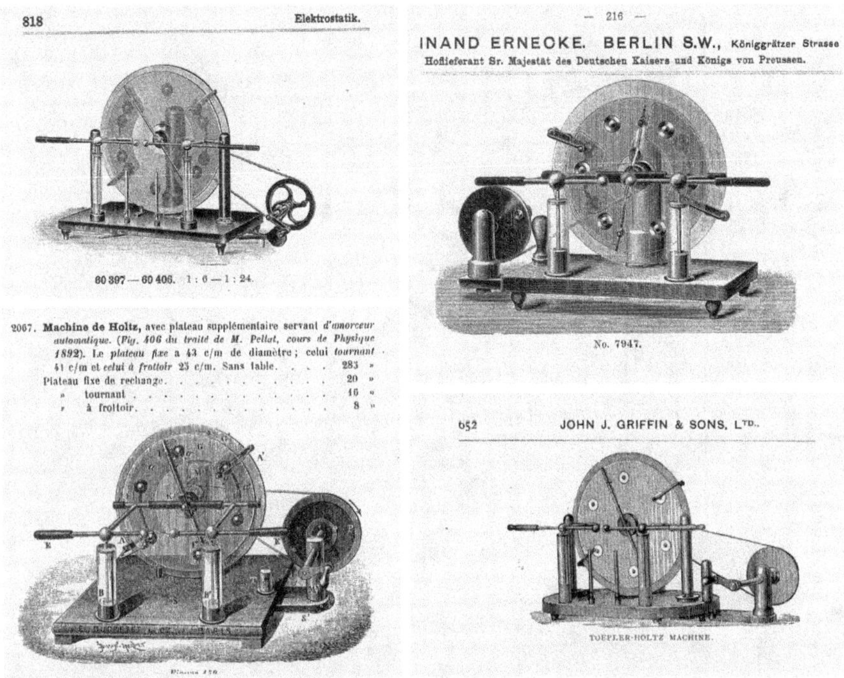

FIGURE 4.5 *The same type of electric influence machine illustrated by four different makers in their catalogues. Max Kohl A.G.,* Preisliste, *Chemnitz, ca. 1911–1912, p. 818; Ferdinand Ernecke,* Preisliste N.18, *Berlin, ca. 1905, p. 216; E. Ducretet et L. Lejeune,* Catalogue . . ., *Paris, 1892, p. 9, and John Griffin and Sons Ltd.,* An illustrated and descriptive catalogue . . ., *London, ca. 1912, p. 652.*

and the English firms of John Griffin and Son Ltd.,[23] and Baird and Tatlock Ltd.,[24] both based in London. These makers were among the most important European suppliers of teaching and research physics instruments of the time. I only selected apparatus whose description is accompanied by an illustration (*Fig. 4.5*). These were often just models for the instruments to be produced but we can assume that the instruments sold or retailed generally corresponded quite closely to the illustration depicting it in the catalogue.

23 John Griffin and Son, Ltd., *Scientific Handicraft An illustrated and descriptive catalogue of Scientific Apparatus*, London, 14th edition, [c. 1912].

24 Baird and Tatlock (London) Ltd., *Price List of Apparatus for Experiments in Practical Physics*, London, 1912.

TABLE 4.6 *The prices of a series of instruments sold by three German makers, a French maker, and two English makers. Prices are given in the original currency and for comparison they are translated into Marks (rounded to the unit). The missing prices indicate that the corresponding maker was not making (or selling) the instrument or proposed a quite different version of it*

	Max Kohl Chemnitz	*Leybold's N. Köln*	*Ernecke Berlin*	*Ducretet Paris*		*John Griffin & Son London*		*Baird & Tatlock London*	
	Mk	*Mk*	*Mk*	*Fr*	*Mk*	*£*	*Mk*	*£*	*Mk*
Mach pendulum	30	30	30			2. 10s	51		
Fessel double gyroscope	50		50			4. 15s	97	4.	82
Mach wave machine	100	100	95			6. 6s	135	6. 10s	133
Morin fall apparatus	360			450	365	18. 18s	386		
Geryk 2-barreled vacuum pump	450	450	450	680	551	20.	408	20.	409
Fountain in vacuum	20	20	18	30	24	1. 4s	25		
Cailltet compression apparatus (300 atm)	650	650		600	486			38.	776
Berthelot apparatus for latent heat	65	60	54	60	49	3. 7s 6d	69	3. 3s	64
Callendar apparatus for the mechanical equivalent of heat	320	320				14. 10s	317		
Carré ice making machine	220	220		250	203	13. 13s	278		

	Max Kohl Chemnitz	*Leybold's N. Köln*	*Ernecke Berlin*	*Ducretet Paris*		*John Griffin & Son London*		*Baird & Tatlock London*	
	Mk	*Mk*	*Mk*	*Fr*	*Mk*	*£*	*Mk*	*£*	*Mk*
Breguet metallic thermometer	46	46	42			2. 12s 6d	54	2. 8s	49
Daniell hygrometer	15	15	18	40	32	2. 15s	56		
Helmholtz double siren	320	320	300	500	405	14.	286	15. 10s	317
19 Helmholtz resonators	150	150		180	146			9. 10s	194
Helmholtz analyser (14 resonators)	600	600	575	700	567			33. 15s	690
Silbermann reflection and refraction apparatus	160	160		200	162	10.	204	7. 15s	158
Lummer-Brodhun photometer	100	160	125			8.	163	7. 17s	160
Weinhold apparatus for showing achromatism	48	48	45			2. 18s	59	2. 18s	59
Becquerel phosphoroscope	120	120	105	135	109	4. 10s	92	5. 10s	112
Direct vision pocket spectroscope	33	33	30	40	32	1. 7s 6d	28	1. 13s	34
Pulfrich refractometer		325	300	605	490	17. 10s	357		
Violle actinometer		550		435	352				

TABLE 4.6 *The prices of a series of instruments sold by three German makers* (cont.)

	Max Kohl Chemnitz	*Leybold's N. Köln*	*Ernecke Berlin*	*Ducretet Paris*		*John Griffin & Son London*		*Baird & Tatlock London*	
	Mk	*Mk*	*Mk*	*Fr*	*Mk*	*£*	*Mk*	*£*	*Mk*
Riess thermometer	40	40	45	175	142	2. 10s		1. 18s	39
Toepler-Voss electrical machine (disk 42 cm)	145	145	145	130	105			7. 10s	153
Wimshurst machine (disk c. 50 cm)	220	190	205	230	186			7. 5s	148
Braun electrometer	56	56				3. 15s	77	3. 15s	77
Sine-tangent galvanometer (Siemens pattern)	250	250	240			16. 16s	343	12.	245
Ruhmkorff-Faraday electromagnet	500		360	850	689	48.	981	30.	613
Demonstration galvanometer	105	120				8. 10s	174		
Desprez-D'Arsonval galvanometer	150							6. 10s	133
Induction coil (25 cm spark)	480	480	405			26.	531	32.	654
Simple Crookes radiometer	5	5	5	10	8	4s 3d	5	6s	6

Certainly the list reported in *Table 4.6*, with about thirty different instruments is too small to be useful for a true statistical study, nevertheless it enables us to draw some interesting conclusions:

a) The two German giants of didactic and laboratory apparatus, Max Kohl of Chemnitz and Leybold's Nachfolger of Köln, proposed very similar instruments, and most of them showed a remarkable uniformity of prices. It is possible that they colluded to maintain identical prices. Furthermore, by carefully comparing these two German catalogues with the above-mentioned French and English ones (and other catalogues not explicitly reported here), it clearly appears that the former advertised a much larger number of instruments related to every branch of physics, optics, and electricity. Most of these instruments were manufactured directly by them, but others (such as specialized optical apparatus, refractometers, saccharimeters, etc.) were made by more focused firms such as Zeiss in Jena, Franz Schmidt und Haensch of Berlin, and then retailed.

b) The prices of a third important German manufacturer Ferdinand Ernecke of Berlin were slightly lower (around 10%) than the ones offered by Max Kohl and Leybold. Ernecke's instruments were similar to theirs, but the firm advertised a more limited range of apparatus.

c) The Parisian firm of Ducretet was by 1900 one of the largest Parisian manufacturers of instruments. Its catalogues include a large number of items, and its prices seemed to vary without a definite trend compared to those of other makers. Some of Ducretet's prices are higher than the German prices; others lower. But Ducretet, like many other French makers was quite conservative and tended to promote instruments conceived in France or of typical French design. They were less open, compared to German or English makers, to adopt or copy apparatus made or invented abroad, which rarely appeared in French catalogues.

d) The prices of the two English makers under consideration, Griffin and Baird & Tatlock, were on average a bit higher than those of their German and French competitors.

e) In a few cases the price differences of two firms were quite significant. Two cases illustrate this: the Pulfrich refractometer sold by Ducretet was around fifty percent more expensive than the one proposed by Max Kohl; the Ruhmkorff-Faraday electromagnet listed by J. Griffin and Son cost almost double the same apparatus in Max Kohl's catalogue. In the first example, the refractometer, which was not made in France,[25] had to be imported from Germany, entailing the usual extra fees, such as import duty, packing, shipping, etc. In the second case, although the catalogues' images were very similar, it was possible that Griffin was selling a larger and more powerful electromagnet. The dimensions were not given in the description of this apparatus. Often differences in size and in design did not appear in the catalogues, which reproduced over and over the same and sometimes old-fashioned engravings.[26] These possibilities might explain some important differences in the prices of instruments made by different firms.

f) Finally, I did not examine the variations of wages of specialized workers in different countries, but I suspect that this factor did not play a fundamental role in the variation of the prices of European instruments.

Therefore, around 1890–1910, European prices of the most common physics instruments were fairly homogeneous. At the time German makers adopted a very aggressive and efficient marketing strategy at the expense of French makers. The latter were losing the predominance, which for several decades they had maintained in the continental market of physical apparatus. But as far as we can see, this trend was not caused by price differences between the French and German manufacturers. The startling success of the Germans at the beginning of the twentieth century was certainly more determined by their innovative design, the quality and variety of their products, and an efficient selling policy of the makers,[27] rather than by purely economic reasons. The prices of instruments made by the two English makers considered were a bit higher than the others. On the Continent English physics instruments were less common than the French ones, which during the nineteenth century were more valued in Italy, Portugal, Spain, Belgium and in several other countries where

25 Instrument makers, and particularly the French ones, were not very keen in reproducing instruments, invented or patented abroad. They often preferred to produce and sell instruments which had been conceived by local scientists and inventors, with whom they often had a good personal relationship. German Pulfrich's and Abbe's refractometers, which were excellent and successful instruments, were not made by Ducretet and not even by the firm Pellin (successor of Duboscq), which at the time was the leading maker of optical instruments and was proposing several other French refractometers. If required, Pulfrich's and Abbe's refractometers had to be imported. See Ph. Pellin, *Instruments d'optique et de précision, Appareil de mesure, VII Fascicule*, Paris, [c. 1901].

26 Catalogues illustrated with photographs became common only after World War I.

27 Large multilingual and profusely illustrated catalogues, and a strong presence at the universal exhibitions were two of the weapons of the clever marketing strategy of the German instrument makers.

the precision industry was not particularly developed. Nevertheless in the late nineteenth century and even more in the beginning of the twentieth century the best English makers whose products were particularly innovative and original were imported to continental Europe, independently of their prices.

European Instruments Imported into the United States

Here I compare the prices of some European instruments sold in Europe with the prices of the same instruments exported to and sold in the United States. In the second half of the nineteenth century, the United States was still importing a large number of instruments from Europe, even if a domestic industry was increasing the number of locally-produced instruments available.[28] Among the imported instruments were sophisticated physics and optical apparatus as well as electrical-measurement instruments.

A comparison between American and European instruments is more difficult if not impossible. For example, if we analyse one of the catalogues of Edward S. Ritchie (1814–1895), one of the most original and ingenious American makers of the nineteenth century, it clearly appears that most of his important instruments (electrical machines, induction coils, vacuum pumps, etc.) were designed very differently from corresponding European instruments. As I pointed out before, different design meant different construction and therefore a possible difference in price. And in fact Ritchie stated in his catalogue's preface that "The instruments enumerated in this Catalogue are almost exclusively of our own manufacture".[29] But in the section dedicated to optics, he specified, "We import from Paris, from the best makers, the following and other Optical Instruments which we cannot make at the cost of importation. The prices will depend on the value of gold, and will be given when desired, by mail".[30] Then he listed a series of optical instruments (most of them from the optician Duboscq) without giving their prices.

28 About the import of instruments to the United States in the late eighteenth century and in the first half of the nineteenth see for example: Alice Walters, "Importing science in the early republic: Union college's 'first purchase' of instruments and books", *Rittenhouse* 16, no. 2 (2002), pp. 85–107, and Deborah J. Warner, "French instruments in the United States", *Rittenhouse* 8, no. 1 (1993), pp. 1–32.

29 E. S. Ritchie and Sons, *Ritchie's illustrated catalogue of philosophical instruments and school manuals*, Boston, 1870, p. iv.

30 Ibid., p. 50.

On the other hand, the firm of James W. Queen and Company of Philadelphia (founded in 1853), certainly one of the most important makers and retailers of scientific instruments in the United States, was importing a large amount of European apparatus in the final decades of the nineteenth century. Many illustrations in Queen's catalogues were copied from European catalogues, and not only because it was easier and cheaper to copy the engravings instead of making new ones, but certainly also because most of these illustrated instruments (and especially the most sophisticated ones) were of European manufacture. This enabled me to compare prices published in two Queen's catalogues[31] with those found in various European catalogues illustrating exactly the same instruments. Among the leading European makers of the 1880s, I selected the French firms of Jules Duboscq (instruments for physical optics),[32] Eugène Ducretet (physics and electrical instruments),[33] and Rudolph Koenig (acoustics apparatus),[34] the English firm Elliott Brothers (electrical and telegraphic instruments),[35] and the German company Hartmann und Braun (electrical measuring instruments).[36] All these firms manufactured high quality apparatus, which was also largely exported.

Before analysing the data we have to remember that scientific instruments imported into the United States from the late eighteenth century onward were subject to an import duty. This duty varied in the course of the nineteenth century, and in the 1880s it was equal to 40% of the value of the imported instrument.[37] We might imagine that in Queen's catalogue the prices included this tax. On the other hand, the American maker clearly specified that "By authority of Act of Congress, June 22, 1874 all Colleges, Schools, Literary Scientific, or Religious Society of the United States are permitted to import, free of duty, Books, Charts, Engravings, and Instruments to be used in connection with the educational exercises of the institution for which they are

31 James W. Queen & Co., *Priced and Illustrated Catalogue of Physical Instruments Chemical Apparatus and Chemicals*, Philadelphia, 1888; and James W. Queen & Co., *Catalogue and Price-List Electrical Testing Apparatus*, Philadelphia, 1887. The most important Queen's catalogues are reprinted in: James W. Queen & Co., *The Queen Catalogues* (with an introduction by D. J. Warner), San Francisco, 1993, 2 vols.

32 Maison Jules Duboscq, *Historique et catalogue de tous les instruments d'optique supérieure appliqués aux sciences et à l'industrie*, Paris, 1885.

33 Eugène Ducretet et C.ie, *Catalogue des instruments de précision, I, II et III parties*, Paris 1879.

34 R. Koenig, *Catalogue des Appareils D'Acoustique, Paris*, 1889.

35 Elliott Brothers, *Catalogue No. E.4*, London, 1895.

36 Hartmann und Braun, *Vollständiges Verzeichnis*, Bockenheim-Frankfurt a. M., 1894.

37 See: *Revised Status of the United States*, Washington DC, 1875, p. 483.

TABLE 4.7 *James W. Queen and Company's published prices compared with those of five European firms. From left to right: Name of the instrument, Queen's price in US Dollars, European maker, Price in European currency, European price converted into US Dollars, Difference between the two prices, Percentage difference*

Instrument	*Queen, 1888 ($)*	*European maker*	*Price*	*Price ($)*	*Diff.*	*Diff. (%)*
Duboscq arc lamp	95	Duboscq, 1885	250 Fr	48	47	98
Silbermann heliostat	250	Duboscq, 1885	600 Fr	114	136	119
Wheatstone photometer	12.5	Duboscq, 1885	35 Fr	7	5.5	78
Balloon for iodine vapours	7	Duboscq, 1885	25 Fr	5	2	47
Soleil saccharimeter	130	Duboscq, 1885	260 Fr	49	81	163
Bohnenberger electroscope	48	Ducretet, 1879	120 Fr	23	25	111
Oersted piezometer	22.5	Ducretet, 1879	90 Fr	17	5.5	32
Breguet thermometer	25	Ducretet, 1879	100 Fr	19	6	32
Eddy current appartus	35	Ducretet, 1879	85 Fr	16	19	117
Daniell hygrometer	16	Ducretet, 1879	40 Fr	8	8	111
Crooke radiometer	5	Ducretet, 1879	12 Fr	2	3	119
Helmholtz double siren	100	Koenig, 1889	450 Fr	86	14	16
Lissajous optical comparator	36	Koenig, 1889	140 Fr	27	9	35
Set of 10 resonators	30	Koenig, 1889	110 Fr	21	9	44
Cagniard de la Tour siren	27	Koenig, 1889	90 Fr	17	10	58

TABLE 4.7 *James W. Queen and Company's published prices* (cont.)

Instrument	*Queen, 1888 ($)*	*European maker*	*Price*	*Price ($)*	*Diff.*	*Diff. (%)*
Sound analyser with 14 resonators	133	Koenig, 1889	650 Fr	124	9	8
Set of 4 tuning forks	32	Koenig, 1889	145 Fr	28	4	16
Large resistance coil box	384	Elliott Bros., 1895	£ 45	234	150	64
Gaugain tangent galvanometer	60	Elliott Bros., 1895	£ 7 10s	44	16	36
Thomson marine galvanometer	220	Elliott Bros., 1895	£ 24	141	79	56
Large set of resistances	384	Elliott Bros., 1895	£ 45	264	120	45
Post office Wheatstone bridge	208	Elliott Bros., 1895	£ 24	141	67	48
Kew magnetometer	520	Elliott Bros., 1895	£ 65	382	138	36
Reading telescope	176	Hartmann & Braun, 1894	440 Mk	106	70	67
Kohlrausch mirror galvanometer	84	Hartmann & Braun, 1894	200 Mk	48	36	75
Kohlrausch universal bridge	52	Hartmann & Braun, 1894	130 Mk	31	21	67
Gaugain tangent galvanometer	60	Hartmann & Braun, 1894	200 Mk	48	12	25

ordered".[38] And a few pages later, Queen's catalogue stated, "We import at the catalogue prices of foreign manufacturers".[39]

Again, the small number of selected cases certainly limits the statistical value of *Table 4.7*. However, it is possible to draw some conclusions from the data:

a) A first (and unsurprising) fact is that instruments imported into the United States (or copied from European models?) were generally much more expensive than those sold in Europe.

b) If we believe that Queen was importing at the prices advertised in European catalogues, the firm's price increases could have been due to the import duty, packing, shipping, insurance, and all the possible expenses related to the transport.

c) If we admit that the import duty of 40% was included in Queen's catalogue prices, some imported instruments, if de-taxed, were cheaper than in Europe. That really is unlikely.

d) If the points b) and c) are true, it is difficult to understand why the price differences varied between 163% for a Duboscq saccharimeter and 8% for a Koenig fourteen-resonator sound analyser. Was the latter made by Queen? But both instruments were complex and typical products of the renowned Paris-based makers, and it is hard to believe that the analyser, whose American price in the United States differed very little from the original Koenig price was made in Boston. If we consider only Koenig's acoustical instruments, how can we explain the discrepancy between a difference of 8% for the analyser and almost 60% for a sound siren?

e) If we consider the six Ducretet instruments, we note that, for four of them, the American prices were around 110–120% higher while the increase for the other two (that is, the piezometer and metallic thermometer) was only 32%. Were the latter manufactured in the United States? In this case the differences in price could be perhaps ascribed to the wages which were higher in the United States than in Europe.

f) The variations in the American prices for the Elliott instruments examined were confined to a relatively narrow interval ranging from 35% to 56%. And the variations of the Hartmann und Braun instruments remained

38 See: Queen, *Priced and Illustrated Catalogue*..., op. cit. (n. 31), inside cover. In fact the import free of duty for educational purposes was applied much earlier than 1874, but probably Queen and other makers did not know that. The complex story of the import duty on scientific instruments during the nineteenth century is retraced in Louis. F. Drummeter Jr., "Philosophical instruments: free of duty! Or: an amateur's foray into legal records", *Rittenhouse* 3, No. 4, (1989), pp. 113–119.

39 Queen, *Priced and Illustrated Catalogue*..., op. cit. (n. 31), "Preface", unnumbered p.

HARTMANN & BRAUN, BOCKENHEIM-FRANKFURT A. M.

No. 391. Universal-Messbrücke nach Kohlrausch.

No. 391. Die **Universal-Messbrücke** nach Kohlrausch unterscheidet sich vor anderen Widerstands-Messapparaten sehr vorteilhaft dadurch, dass die Widerstände direkt in Ohm an einer Skale ohne Zuhilfenahme einer Tabelle abgelesen werden können. Der Rheostat enthält Vergleichswiderstände von 1, 10, 100 u. 1000 Ohm.*) In Verbindung mit einem geeigneten Galvanometer (etwa No. 366, Seite 7, oder No. 536a, Seite 15), dient der Apparat zur Widerstandsmessung fester Leiter, z. B. von Drähten von 0.1 bis 10000 Ohm, mit einer für technische Zwecke ausreichenden Genauigkeit. Unter Verwendung von Wechselströmen, die durch ein kleines am Apparat befindliches Induktorium erzeugt werden, und unter Ersetzung des Galvanometers durch ein geeignetes Telephon, können auch Widerstände von Elektrolyten, z. B. der innere Widerstand von Elementen, sowie die Übergangswiderstände von Erdplatten an Blitzableiter-Anlagen bestimmt werden. Preis M. 130.–

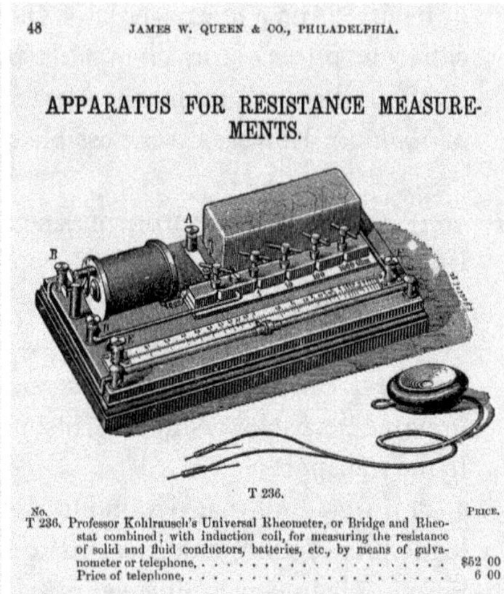

48 JAMES W. QUEEN & CO., PHILADELPHIA.

APPARATUS FOR RESISTANCE MEASUREMENTS.

T 236.

No.		PRICE.
T 236.	Professor Kohlrausch's Universal Rheometer, or Bridge and Rheostat combined; with induction coil, for measuring the resistance of solid and fluid conductors, batteries, etc., by means of galvanometer or telephone,	$52 00
	Price of telephone,	6 00

FIGURE 4.6 *Kohlrausch universal bridge in the catalogue of Hartmann und Braun* (Vollständiges Verzeichnis, *Bockenheim-Frankfurt a/M, 1894, p. 60), and the same instrument proposed in the United States by James W. Queen and Co. in 1887.* (Catalogue…, *Philadelphia, 1887, p. 48*).

between 67% and 75%, apart from the Gaugain galvanometer, which was only 25% more expensive in the United States (*Fig. 4.6*).

In fact, *Table 4.7* raises a lot of questions. Why did the differences in price vary so much? Perhaps Queen was able to negotiate different wholesale prices with the French, German, or English makers. Or the cost of packing, shipping, insurance were really different for different European countries. Furthermore, how can we explain the huge differences in the percentage of price mark-ups for instruments by French makers? It has been suggested that perhaps Queen set different prices based on the volume of expected sales.[40] For example, the price of a Soleil saccharimeter, of which only a few were sold, was kept higher to cover the fixed costs, while it was possible that a much more 'popular' Gaugain galvanometer was sold at a proportionally lower price. Finally, we have to point out that we are not absolutely sure that all of the above-mentioned instruments were actually imported. I am convinced that most of them were, but there is something interesting in the Queen catalogue. In the section dedicated to induction coils we can read: "The list comprises

40 Richard Kremer, private communication.

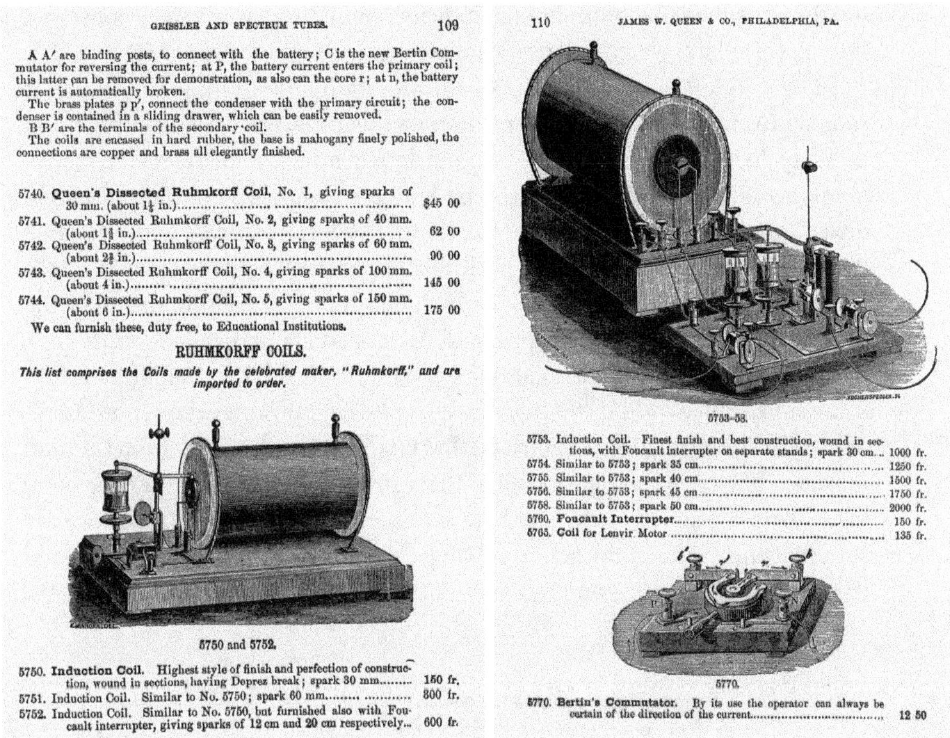

GEISSLER AND SPECTRUM TUBES. 109

A A′ are binding posts, to connect with the battery; C is the new Bertin Commutator for reversing the current; at P, the battery current enters the primary coil; this latter can be removed for demonstration, as also can the core r; at n, the battery current is automatically broken.

The brass plates p p′, connect the condenser with the primary circuit; the condenser is contained in a sliding drawer, which can be easily removed.

B B′ are the terminals of the secondary coil.

The coils are encased in hard rubber, the base is mahogany finely polished, the connections are copper and brass all elegantly finished.

5740. **Queen's Dissected Ruhmkorff Coil**, No. 1, giving sparks of 30 mm. (about 1¼ in.) $45 00
5741. Queen's Dissected Ruhmkorff Coil, No. 2, giving sparks of 40 mm. (about 1⅝ in.) 62 00
5742. Queen's Dissected Ruhmkorff Coil, No. 3, giving sparks of 60 mm. (about 2⅜ in.) 90 00
5743. Queen's Dissected Ruhmkorff Coil, No. 4, giving sparks of 100 mm. (about 4 in.) 145 00
5744. Queen's Dissected Ruhmkorff Coil, No. 5, giving sparks of 150 mm. (about 6 in.) 175 00

We can furnish these, duty free, to Educational Institutions.

RUHMKORFF COILS.

This list comprises the Coils made by the celebrated maker, "Ruhmkorff," and are imported to order.

5750 and 5752.

5750. **Induction Coil.** Highest style of finish and perfection of construction, wound in sections, having Deprez break; spark 30 mm. .... 150 fr.
5751. Induction Coil. Similar to No. 5750; spark 60 mm. .... 300 fr.
5752. Induction Coil. Similar to No. 5750, but furnished also with Foucault interrupter, giving sparks of 12 cm and 20 cm respectively... 600 fr.

110 JAMES W. QUEEN & CO., PHILADELPHIA, PA.

5753–58.

5753. Induction Coil. Finest finish and best construction, wound in sections, with Foucault interrupter on separate stands; spark 30 cm. .. 1000 fr.
5754. Similar to 5753; spark 35 cm .... 1250 fr.
5755. Similar to 5753; spark 40 cm .... 1500 fr.
5756. Similar to 5753; spark 45 cm .... 1750 fr.
5758. Similar to 5753; spark 50 cm .... 2000 fr.
5760. **Foucault Interrupter** .... 150 fr.
5765. **Coil for Lenvir Motor** .... 135 fr.

5770.

5770. **Bertin's Commutator.** By its use the operator can always be certain of the direction of the current 12 50

FIGURE 4.7 *Two pages of the catalogue of James W. Queen and Co. of 1888* (Priced and Illustrated Catalogue…, *Philadelphia, 1888, pp. 108–109*) *illustrating various induction coils. The prices of the coils made by the American maker are in Dollars, while those of the French imported coils are in Francs.*

the coils made by the celebrated maker 'Ruhmkorff' and are imported to order".[41] But it is quite strange to see that the prices of the imported French induction coils (from 30 mm to 500 mm sparks) were left in French Francs in the American catalogue. We do not find any other instrument listed in it with French prices. And these prices perfectly correspond to the ones advertised in the catalogue of various French makers (*Fig. 4.7*).[42] Why is that? Because the coils were imported only by special order? Does this suggest that the

41 Queen, *Priced and Illustrated Catalogue…*, op. cit. (n. 31), pp. 109–110.

42 By this time Rumkhorff was dead, but its production was continued by his successor Carpentier, and also Ducretet was proposing Ruhmkorff coils in his catalogues. See Ducretet, op. cit. (n. 33), pp. 106–107.

instruments made by other foreign makers (and whose prices were converted into U. S. Dollars) were imported regularly and not by special order?

Table 4.8 gives some examples of the situation around 1910–1912, which is roughly twenty-five to thirty years later. By this time, the American precision industry had made much progress, and larger numbers of scientific instruments were being made in the United States. I compare here the prices of a dozen instruments illustrated in Max Kohl's catalogue No. 50[43] with the same instruments offered by the firm L. E. Knott Apparatus Company of Boston.[44] In the preface of Knott's catalogue, it was stated that "A large proportion of the articles offered have been designed and developed in our own laboratory [...]. In addition we have made a careful selection from the best scientific establishments of the world".[45] So there always was a significant number of instruments (mostly optical and electrical) that were imported, mostly from German makers. I have to point out that all of these instruments in the American catalogue were captioned "duty free" before the price in U.S. Dollars. Thus, we are sure that they were imported—probably from the firm Max Kohl, which sold the same instruments—and not manufactured in the United States.

TABLE 4.8 *Comparison of prices around 1910–1912. From left to right: Name of the instrument, Max Kohl's prices in Marks, Prices in Marks converted into US Dollars, Prices of the instrument sold by Knott, Difference between the two prices, Percentage difference*

Instrument	*M. Kohl (Mk)*	*$*	*Knott ($)*	*Diff.*	*Diff. %*
Galton whistle	55	13	16.5	3.5	*27*
Double-barreled Geryk vacuum pump	350	84	95	11	*13*
Weinhold demonstration prism	48	12	17.25	5.25	*44*
Hefner standard lamp	46	11	16.5	5.5	*50*
Lummer-Brodhun photometer	125	30	39	9	*30*

43 See Max Kohl AG, op. cit (n. 19).

44 L. E. Knott Apparatus Company, *A catalogue of physical instruments Catalogue 17*, Boston, 1912.

45 Ibid., "Preface", unnumbered p.

Instrument	*M. Kohl (Mk)*	*$*	*Knott ($)*	*Diff.*	*Diff. %*
Dolezalek quadrant electrometer	160	38	55	17	*45*
Toepler machine (3 disks of 30 cm)	55	13	21	8	*59*
Wimshurst machine (2 ebonite disks 50 cm)	140	34	49.5	15.5	*46*
Wiechert quartz filament electrometer	190	46	62	16	*36*
Hydrodynamic model of Wheatstone bridge	80	19	34.5	15.5	*82*
Rotating magnetic field apparatus	120	29	40	11	*39*
Thomson repulsion apparatus	150	36	55	19	*53*

Again, and it is not a surprise, the instruments imported and sold in America were more expensive than those sold in Europe. The average increase in price (around 44%) seems to be lower than the increase of the 1880s. But when the prices were labelled "duty free", did this mean that customers buying for non-educational purposes (for example, for technical use in a factory) had to add an extra 40% for the import duty? If so, the average increase of prices for instruments sold in the United States to non-educational institutions would have been 80% to 90% higher than the original European prices.

Once again, we find it difficult to explain why the price increase for a Geryk pump and that for a "hydrodynamic model" were respectively remarkably lower and higher than the price increases of other apparatus. Certainly the pump was much more common in laboratories and may have been a best-seller compared to the hydrodynamic model of a Wheatstone bridge.[46] Perhaps as we suggested before, the pump could be sold in the United States at a relatively lower price than the model, which was more specialized and less needed piece of apparatus.

46 Having visited a large number of historical collections, I have seen several pumps of this type, but no hydrodynamic model.

But for the moment, all these thoughts are pure speculations. To be on firmer ground, we must consider a larger number of relevant instruments, which would be accessible to statistical analysis. In order to answer to the stated questions and to better understand the trade of instruments between Europe and the United States, we should also use more refined tools. Ideally, we would examine the manufacturers' archives to discover invoices, contracts, and perhaps special agreements between the most important instrument makers and retailers in Europe and in the United States. Unfortunately, as far as I can see, the commercial archives of renowned French makers such as Duboscq, Koenig, and many others do not exist anymore. And in preliminary examination of the Max Kohl archives, no documents related to the company's trade with the United States were found.[47]

Conclusion

In my research I essentially examined the prices of instruments used in physics cabinets and laboratories for teaching, research, and measurements. My main source of information was a collection of illustrated trade catalogues. I did not consider industrial and technical instruments, whose production grew dramatically from the end of nineteenth century onward. The analysis of the data concerning the prices of some instruments made in France during the nineteenth century basically showed their great stability. For several decades the same instruments were advertised with no or very little variation in price. A comparison of several instruments made in France, England, and Germany between the end of the nineteenth century and the first decade of the twentieth shows a remarkable uniformity in the prices of three of the leading German makers, while the prices of the same instruments made by two important English makers tended to be sometimes higher. On the other hand, the prices of the French firm Ducretet exhibited some variations that did not show a specific trend. Finally we can affirm that the choice of an instrument maker as the supplier responsible for equipping any first-class collection was not (or certainly only slightly) determined by the cost of that company's instruments. Other factors, such as the quality of the instruments, the number and variety of those advertised, their immediate availability, the reputation of a firm, the efficiency of its advertising and its service, and its geographical proximity, were certainly more important to the decision making.

47 I have to thank Paul Zoller, who had the opportunity to see the Max Kohl archives and told me in a private communication that he did not see any document of this kind.

Much more problematic is the comparison between the prices of instruments made in Europe and the prices of the same instruments imported and sold in the United States. Certainly the latter were remarkably higher than the former. This can be explained by the duties paid to import such goods into the United States, and by the various fees for packing, shipping, and insuring them. Nevertheless, the increases in prices seem to have been quite erratic. In the cases described above, the prices oscillated between about 15% to more than 150%. I have proposed some possible and reasonable explanations for these huge variations, but they all remain purely speculative. Without further and deeper research many questions will remain unanswered. Certainly it would be necessary to examine a larger collection of trade catalogues and compare the prices of a larger number of instruments. But, even more important would be further inquiry into the commercial manufacturers' archives (which unfortunately have rarely been preserved) in order to search for documents explaining the conditions regulating the prices of instruments made in Europe and sold in the United States.

CHAPTER 5

Mathematical Instruments Changing Hands at World's Fairs, 1851–1904

Peggy Aldrich Kidwell

From the 1851 Crystal Palace exhibition in London to the 1904 Louisiana Purchase Centennial Exposition in St. Louis, Missouri, world's fairs were places where mathematical instruments changed hands. Mathematicians, inventors and entrepreneurs sought customers for their innovations and the recognition of international panels of judges. The instruments on display included early commercial adding and calculating machines, pioneering difference engines, newly designed slide rules, and sophisticated geometric models. Vendors also sold traditional drawing instruments like protractors and rules, often manufactured using new materials and production techniques that greatly reduced cost. These appealed to a broad range of potential customers, including mathematicians, engineers, and scientists. Exhibits at fairs not only contributed to the spread of ideas and products, but were used by governments to demonstrate the modernity and intellectual achievements of nations. Both the mathematical community and people represented at the fairs came to include a much larger portion of the world. Surviving objects, printed documents, and archival materials amply demonstrate the improvement, diffusion, and internationalization of the material culture of mathematics as it occurred through exhibition at the fairs.

A few of the objects shown in the period 1815–1904 illustrate the role of fairs in shaping the exchange of instruments and ideas. As a list of important fairs indicates (*Table 5.1*), the events were held most frequently in London and Paris, with occasional forays elsewhere in Europe and to the United States.

Several things happened—or sometimes happened—to mathematical instruments shown at these expositions. First, they were displayed. A local committee decided the general classification of exhibits and allotment of space. Within this scheme, particular exhibits were designed by the exhibitors, be they individuals, organizations, or government bodies. Decisions about 'who exhibited what' depended on local, national, and international politics as well as the novelty and quality of inventions.

Second, the mathematical objects shown were publicized. Exhibitors published books and brochures and arranged demonstrations. Local, national, and

 | DOI 10.1163/9789004324930_006

TABLE 5.1 *Important World's Fairs, 1851–1904*

1851	London	Great Exhibition of the Works of Industry of All Nations (Crystal Palace)
1853–1854	New York City	Exhibition of the Industry of All Nations
1855	Paris	Exposition universelle
1862	London	International Exhibition of 1862
1867	Paris	Exposition universelle
1873	Vienna	Weltausstellung
1876	Philadelphia	United States Centennial International Exposition
1878	Paris	Exposition universelle
1885	London	International Exhibition of Inventions
1889	Paris	Exposition universelle de 1889
1893	Chicago	World's Columbian Exposition
1900	Paris	Exposition universelle de 1900
1904	St. Louis	Louisiana Purchase Centennial Exposition

international press reports appeared. Official fair publications offered descriptions. Voluminous records of publications survive and are increasingly available to scholars in electronic form.

Third, entries were judged by an international panel of experts—preferably experts who had examined the displays, seen the literature, and witnessed the demonstrations. The judges might grant an award, which could be mentioned in later advertising.

Fourth—in many cases—instruments changed hands. They sometimes were purchased by enthusiastic visitors. They occasionally were collected on the spot as specimens, most often by government agencies. Even those instruments that did not change hands at the fairs often were later collected as a result of publicity they had received. Finally, to a considerable degree, instruments exhibited at fairs have been preserved long past their use in mathematics, science, and engineering. Objects which have found their way into the collections of the Science Museum in London, the Smithsonian Institution in Washington, D.C., and the Conservatoire National des Arts et Métiers in Paris, well illustrate this point. The connection of these objects to fairs has not always been well known. In other words, one fruitful question historians can ask about mathematical instruments from this time period is: What tie might there be between this object and a world's fair? Some inventors were spurred to new work by fairs, visitors made their purchases, and publicity associated with fairs sometimes gave objects lasting prominence.

The simple adding machine designed by the Swiss clockmaker Victor Schilt (*Fig. 5.1*), owes much of its luster to a fair. It was exhibited in London at the Great Exhibition of the Works of Industry of All Nations, otherwise known as the Crystal Palace exhibit of 1851. It was one of the first successful adding machines in which numbers were entered by pressing keys. Such keys are among the few features of adding machines to survive on modern electronic calculators. Schilt's machine was duly described in accounts of the fair, and the judges awarded him an honorable mention. According to reports, several potential customers asked to purchase copies of the machine, but Schilt declined the offers. Denis Roegel has pointed out that Schilt's machine closely resembles one already patented by another Swiss clockmaker, Jean-Baptiste Schwilgué, which may explain this decision. Schilt did keep the machine he had shown. It was later purchased from his heirs by the American inventor of key-driven adding machines and collector Dorr E. Felt, and subsequently donated to the Smithsonian.[1]

A second object exhibited at a world's fair, the difference engine of the Swedes Georg and Edvard Scheutz (*Fig. 5.2*), is better known to historians of computing. As early as the 1820s, the English mathematician and natural philosopher Charles Babbage proposed a machine of this type, which was to calculate tabular values using the method of finite differences and print the results. By 1832, Babbage had completed a working section of his difference engine. He obtained considerable funding from the British government to build the machine, but difficulties with workmen, changes in government, and the prospect of more complex computing devices combined to prevent its completion.

The Swedish lawyer, publisher, journalist, and inventor Georg Scheutz (1785–1873) was so intrigued by accounts of the difference engine that he enlisted his son, the engineer Edvard Scheutz, actually to build such a machine. By 1853, with the aid of skilled Swedish machinists and government funding, they had completed the task and set out to sell the machine. The elder Scheutz had some doubts about Swedish abilities to market inventions. In 1834, after the failure of an industrial exhibition in Stockholm, he had commented:

> A Swedish craftsman who has carried out a good piece of work, often believes that it is enough to have done it. He knows that the product

1 Peggy A. Kidwell and Paul Ceruzzi, *Landmarks in Digital Computing: A Smithsonian Pictorial History*, Washington, D.C., 1994, p. 33; Denis Roegel, "An Early (1844) Key-Driven Adding Machine", *IEEE Annals of the History of Computing* 30 (2008), pp. 59–65.

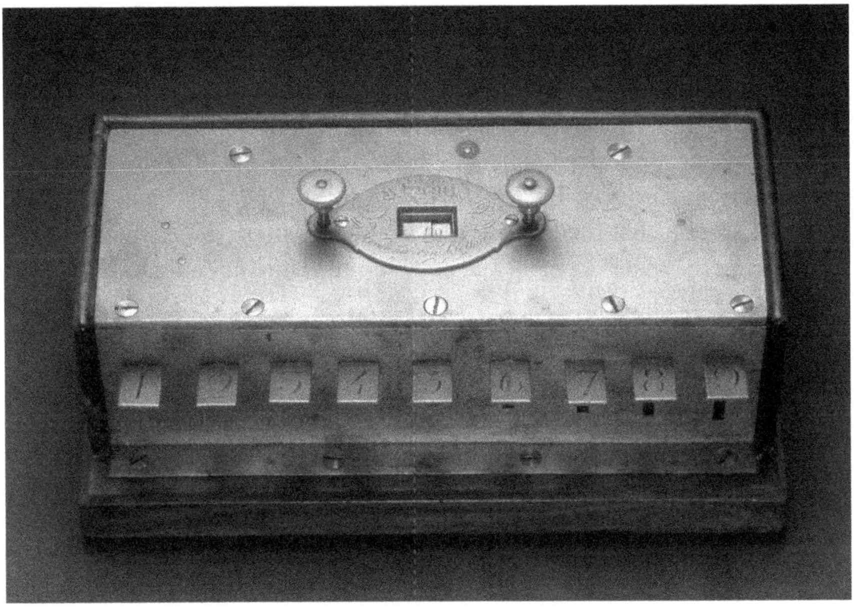

FIGURE 5.1 *Schilt adding machine exhibited at the Crystal Palace in London in 1851.* SMITHSONIAN IMAGE 89-13539.

FIGURE 5.2 *Scheutz difference engine exhibited at the Exposition universelle in Paris in 1855.* SMITHSONIAN IMAGE 89-22028.

> is good, but he often does not care if others know it. Instead of 'bringing his goods to the market' in a purposeful way, so that he receives not only reward for his industry or his ingenuity, but also what is of chief importance, the stimulus to fresh endeavour, he vegetates impotently in the shade, unnoticed, unpaid, often struggling with embarrassment and sometimes even with harsh poverty itself; and the skills he has acquired, the ideas he has put forward, and all the inventions due to him, are often lost to the trade or perish and are forgotten with his passing.[2]

Seeking to avoid this fate, Georg and Edvard Scheutz enlisted the aid of the well-born Swedish engineer Pehr Amjörn Sparre. Sparre agreed to cover expenses that the Scheutzes would incur visiting England to demonstrate their engine there, as well as the cost of displaying the machine at the Paris Exposition of 1855. In return for this cash advance, Sparre was to receive both patent rights and £ 4,000 if the engine sold. On their visit to England, the Scheutzes won the good will of the engineer William Gravatt (1806–1866), who arranged to have the difference engine demonstrated at the Royal Society of London. This event went well, a committee of the Royal Society reported positively on the invention, and Babbage was successfully enlisted to promote the machine. However, no British buyer was found for the Scheutz engine.

In the summer of 1855, as planned, the difference engine was moved to Paris for display at the Exposition universelle. William Gravatt even went along to demonstrate the machine. Babbage came by in October to provide further publicity, suggesting to the Parisian Académie des Sciences how the operation of the difference engine could be represented in a notation he had devised. These efforts successfully garnered one of the Exposition's gold medals for the Scheutzes and their difference engine. Moreover, at the Exposition Babbage met with the American astronomer Benjamin Apthorp Gould (1824–1896). Gould was in Europe to purchase instruments for the U.S. Coast Survey and the newly established Dudley Observatory in Albany, New York.

Gould was impressed by what Babbage showed him. In April of 1856 he wrote to Babbage to inquire if the Scheutz difference engine was for sale, and if so for how much. He also requested more detailed information about what the machine could do. By this time, potential purchasers in both England and France had declined to buy the difference engine. When Gould's inquiry was

---

2 [Georg Scheutz], "Slöjdexpositionen i Stockholm 1834", *Journal för Manufakturer och Hushållning*, June, (1834), p. 214; in Michael Lindgren, *Glory and Failure: The Difference Engines of Johann Müller, Charles Babbage and Georg and Edvard Scheutz* (trans. by Craig G. McKay), Cambridge, 1990, p. 91.

forwarded to the Scheutzes, they agreed to a price of £ 1,000. Thus the Scheutz difference engine became the first difference engine sold. It operated briefly in Albany, but use of it ceased after Gould left the directorship of the Dudley Observatory in early 1859.[3]

The story of difference engines at the fairs did not end with the Scheutz. Babbage and Gravatt arranged to display part of Babbage's incomplete difference engine at the International Exposition of 1862, and the Swede Martin Wiberg exhibited a much smaller difference engine at the same fair. In 1876, the American George B. Grant exhibited a difference engine at the Centennial Exposition in Philadelphia. However, difference engines proved more important as symbols of mechanical prowess than practical products. The focus of both inventors and customers was moving toward smaller, desktop devices. This trend is amply demonstrated by the experiences of the American university president Frederick A. P. Barnard (1809–1889) at the 1867 Exposition universelle in Paris.

Barnard had a lifelong interest in scientific and technical apparatus. A graduate of Yale College, he indulged his interest in instruments as a professor at the University of Alabama, and then as a professor and later chancellor at the University of Mississippi. Barnard left Mississippi at the outbreak of the Civil War, and soon was named President of Columbia College in New York City. In 1866 he was appointed as one of the U.S. Commissioners to the Paris Exposition of 1867. This gave him further opportunities to examine and acquire objects.

The American commissioners in Paris studied different classes of objects in separate groups. Barnard was to examine machinery and processes of the industrial arts. He also judged exhibits of "instruments of precision and apparatus for instruction in science." When some of the American commissioners failed to file their reports, Barnard prepared not only the U.S. report on industrial improvements, but a description of apparatus for the exact sciences.[4]

Barnard's report included a chapter on metrology and mechanical aids to computation. In the third and final section, Barnard turned his attention

---

3 Uta C. Merzbach, *George Scheutz and the First Printing Calculator*, Washington, D.C., 1977, pp. 26–28; Lindgren, op. cit. (n. 2), pp. 202–211 and 282–284; Mary Ann James, *Elites in Conflict: The Antebellum Clash over the Dudley Observatory*, New Brunswick, N.J., 1987, pp. 113–213.

4 Frederick A. P. Barnard, "Machinery and Processes of the Industrial Arts and Apparatus of the Exact Sciences," in William P. Blake (ed.), *Reports of the United States Commissioners to the Paris Universal Exposition*, Washington, D.C., 1870, 6 vols.: v. 3; P. A. Kidwell, "American Scientists and Calculating Machines: From Novelty to Commonplace", *Annals of the History of Computing* 12 (1990), pp. 31–40: 32–33.

to mechanical calculation, especially the arithmometer of Charles Xavier Thomas. Like Thomas, Barnard described the machine as part of a tradition of calculating devices that included the calculating rods of John Napier and Blaise Pascal's adding machine. He included a rather detailed discussion of the principle of operation of difference engines.[5] His survey was one of the first accounts of the history of computing devices published in the United States.

Thomas's calculating machine had been commended as early as the Crystal Palace exhibition of 1851, and received a silver medal from the international jury in Paris in 1867. Barnard also was delighted. As a teacher and arithmetic textbook author, he much appreciated the problems most people had doing calculations, and found it remarkable that a machine could do this work. As he wrote:

> To most persons the process of calculation involves a species of mental labor which is painful and irksome in the highest degree; and to such, no part of their educational experience recalls recollections of severer [sic] trials, or of burdens more difficult to bear. That this toil of pure intelligence—for such it certainly seems to be—can possibly be performed by an unconscious machine is a proposition which is received with incredulity; and even when visibly demonstrated to be true, is a phenomenon which is witnessed with unmingled astonishment.[6]

Barnard described the advantages of the arithmometer minutely, comparing it to similar products made in France and in Russia.

Unsurprisingly, Barnard wanted to have his own arithmometer. In September of 1867 he wrote to the trustees of Columbia requesting a special appropriation to buy instruments for the college. He initially thought $ 1,000 would do, but soon decided he needed twice as much. The trustees complied. Most of Barnard's purchases went to various science departments at Columbia, but the arithmometer was for the President's room. Barnard used his arithmometer in preparing his report on the Exposition of 1867, but the novelty wore off. In 1886, he transferred it to the Columbia College Observatory. From there it went to the Museum of Science and Industry in Chicago and then came to the Smithsonian Institution in about 1966 (*Fig. 5.3*). By that time, the instrument's tie to 1867 fair could only be reconstructed from marks on the object and archival sources.

5 Barnard, op. cit. (n. 4), pp. 629–648.

6 Ibid., p. 629.

FIGURE 5.3 *Thomas arithmometer purchased by an attendee of the Exposition universelle in Paris in 1867.*
SMITHSONIAN IMAGE 87-6436.

At the early world's fairs, exhibitors tended to be companies and individuals. The success of fairs encouraged governments and non-profit institutions to sponsor exhibits as well. One of the largest and most intriguing exhibits at the Centennial Exposition of 1876 included drawing instruments shown by the Japanese Empire Department of Education. After the Meiji restoration of 1868, Japanese leaders had embraced many of the goals and institutions of the Western nations who had forcibly opened the country to international trade. The new government replaced the feudal authorities with a powerful central government. It tolerated the publication of newspapers, which came to have a large circulation. The central government also introduced a conscript army and compulsory public education.

These changes were part of an attempt to establish Japan as a modern world power. Treaties the country had signed in the 1850s with the United States, Britain, France, the Netherlands, and Russia had not only opened the nation to Western traders but set tariffs on both imports and exports. Moreover, Japan had no legal jurisdiction in the eight Japanese ports open to trade. Foreign nationals accused of crimes in Japan were tried in courts presided over by foreign judges under foreign laws. Hoping for more equitable treatment, the

Japanese government sent a delegation of some fifty high officials on a twenty-two month tour of the United States and Europe in the early 1870s. Members of the delegation not only petitioned for the modification of the so-called 'unequal treaties', but made careful studies of Western industry, government, and schools.[7]

Western powers initially dismissed attempts to revise the unequal treaties. The traveling Japanese did learn much about educational practices and also exhibitions. Several Japanese companies showed their wares at a hastily arranged display at the 1873 fair in Vienna. This proved sufficiently successful for the government to prepare a small exhibition open to foreigners in Kyoto, a city that previously had been restricted to Japanese.[8] With ample time to arrange exhibits for the Centennial Exhibition in Philadelphia, the Meiji authorities threw themselves into the task, spending money to assemble collections, advancing funds to manufacturers, and paying the cost of transport to Philadelphia of both goods and exhibitors. The Imperial Commissioners also published in English a 130-page official catalogue of the Japanese exhibit.[9]

The protractor and set of rules shown in *Fig. 5.4* and *Fig. 5.5* were among the objects chosen by the Japanese Department of Education to show what was new and Western in Japan. Also exhibited were printed materials, including *An Outline History of Japanese Education* prepared especially for the exhibition.[10] Other documents displayed were laws relating to schools; reports of the Department of Education; regulations of government schools; examination papers; accounts of libraries, museums, and botanical gardens; and a wide range of books, maps, academic journals, and newspapers. The judges at the Centennial Exhibition were much impressed by the Japanese educational efforts. They commended the Department of Education generally for "A full

---

7 For introductions to this period in Japanese history, see Marius B. Jansen, *The Making of Modern Japan*, Cambridge, MA, 2000; Andrew Gordon, *A Modern History of Japan, from Tokugawa times to the Present*, Oxford, 2003. For an overview of the introduction of Western mathematics into Japan, see Chikara Sasaki, "The Emergence of the Japanese Mathematical Community in the Modern Western Style, 1855–1945", in Karen Hunger Parshall and Adrien C. Rice (eds.), *Mathematics Unbound: The Evolution of an International Mathematical Research Community, 1800–1945*, Providence and London, 2000, pp. 229–252.

8 Samuel Mossman, *New Japan, the Land of the Rising Sun*, London, 1873, pp. 449–452.

9 Imperial Japanese Commission to the International Exhibition at Philadelphia, *Official Catalogue of the Japanese Section, and Descriptive Notes on the Industry and Agriculture of Japan*, Philadelphia, 1876, pp. 3–4.

10 Japan. Japanese Department of Education, *An Outline History of Japanese Education*, New York, 1876.

FIGURE 5.4 *Japanese circular protractor exhibited at the Centennial International Exposition in Philadelphia in 1876.*
SMITHSONIAN IMAGES 2003-35924.

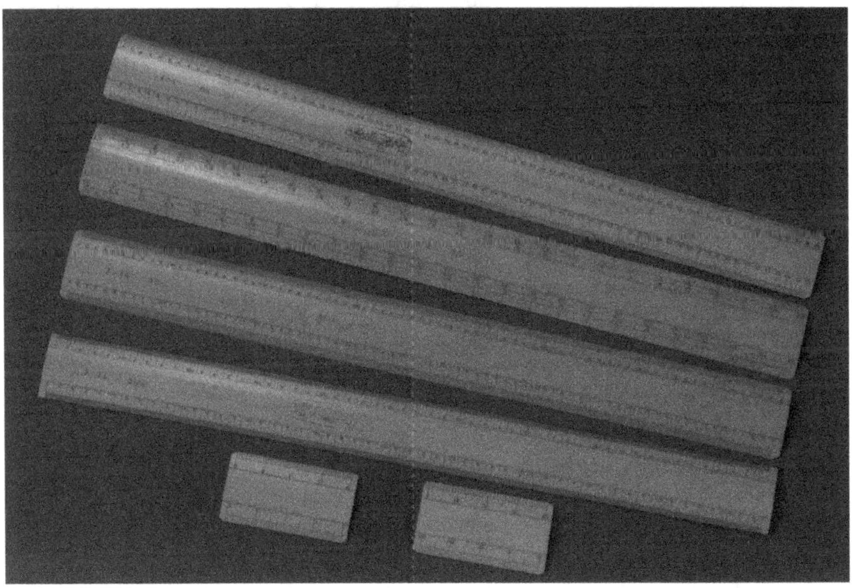

FIGURE 5.5 *Japanese set of rules exhibited at the Centennial International Exposition in Philadelphia in 1876.*
SMITHSONIAN IMAGES 2003-35926.

exhibit of the recent progress of the empire of Japan in the promotion of public education, showing enlightened aims and creditable progress".[11] Ancient and modern surgical and medical exhibits from the Medical Academy of Tokyo; zoological and botanical specimens from the Tokyo Museum; and apparatus, drawings, reports and photographs from the Imperial University in Tokyo all were recognized. The drawing instruments, rightly enough, were not singled out for attention.[12]

John Eaton, the United States Commissioner of Education, had long believed that exhibits could benefit educators. He not only played an active role in preparing a United States government exhibit on American schools at the Philadelphia world's fair, but in addition encouraged states and cities to prepare their own exhibits relating to schools. Eaton also saw the fair as an opportunity to advance a cause that he had advocated since at least 1870, the establishment of a national educational museum at the Bureau of Education. Eaton saw exhibits like that of the Japanese as possible acquisitions for his planned museum. When the Japanese vice-minister of education visited Washington, Eaton persuaded him to arrange the donation of "the entire education exhibit of Japan" to the Bureau of Education "on behalf of education in the United States".[13]

The Japanese attempt to demonstrate the modernity of their nation succeeded. In the 1890s, several of the most objectionable provisions of the treaties Japan had signed with Western nations were revised to grant Japan greater governance over setting tariffs and legal control over foreign nationals on its soil.[14] It is not clear whether the Japanese instruments ever were actually displayed in Washington. They were catalogued in 1887, and a small group of accumulated objects was put on display. In 1910 the remnants of the collection were transferred to the Smithsonian Institution, and were dispersed throughout collections in anthropology, biology, and geology. With changes in the organization of the museum over time, some of these materials eventually made their way to the mathematics collections. Recovering the connection between these drawing instruments and the 1876 fair took considerable research.[15]

11 Francis A. Walker (ed.), *International Exhibition, 1876. Reports and Awards*, Washington, D.C., 1880–1884, 11 vols.; v. 8, p. 335.

12 Ibid.

13 U.S. Bureau of Education, *Annual Report of the Commissioner*, Washington, D.C., 1876, pp. ccxi–ccxii.

14 Gordon, op. cit. (n. 6), pp. 91–92 and 119; Jansen, op. cit. (n. 6), pp. 423–429.

15 National Museum of American History (NMAH), Smithsonian Institution, Washington, D.C., Collection Records, Accession files 51115 and 51116. An examination of the *Annual*

Slide rules, like drawing instruments, had been used long before the era of world's fairs. Some were exhibited in London in 1851 and in Paris in 1867. The International Inventions Exhibition held in London in 1885 played a larger role in diffusing knowledge of the instrument in the United States. The linear slide rule invented by French general Amadée Mannheim in the mid-nineteenth century, as well as cylindrical slide rules of the Irish engineering professor George Fuller and the American 'computing engineer' Edwin Thacher were particularly influential. Both Fuller's and Thacher's rules were initially manufactured by the English firm of Stanley, and were exhibited in London in 1885 among that company's products.[16] In the next decade, the handheld, linear slide rule, based on Mannheim's design, spread throughout England and the United States.[17] Instruction in the slide rule joined experiments with far larger machines in the laboratories of engineering schools. Improved materials, particularly celluloid, made it possible to make relatively inexpensive, sturdy slide rules that were easy to read. British, German, and American firms soon were manufacturing Mannheim and other linear slide rules. While the use of the instrument did not become universal, it did become a symbol of the calculating community.[18]

In 1889, fair goers went to Paris to mark the centennial of the French Revolution, admire the newly-built Eiffel Tower, and see the best and the latest mathematical instruments. These included the tabulating machine exhibited by Herman Hollerith of Washington. Hollerith's machine counted data

---

Reports of the Commissioner of the Bureau of Education and a comparison of the published listings of items exhibited by the Japanese in Philadelphia with surviving object make the account given above seem highly likely. One sliding rule even was received with a label indicating that it was associated with the Japanese Empire Department of Education. Another piece of physical apparatus, which may be an orrery, is specifically marked as a product of the "Kaisei school factory" in Tokyo, which is linked to the exhibit at the fair. A second orrery is dated December, 1875 and specifically said to be a copy of an English instrument made in Tokyo. I thank Steven Turner of NMAH for information about the orreries.

16 W. F. Stanley received a gold Medal at the fair for "improvements in philosophical instrument." See: International Inventions Exhibition, *The Awards of the International Juries, Confirmed and Issued by the Jury Commissioners*, London, 1885, p. 32.

17 "Gravet's Calculating Rule," *Engineering* 44 (1887), p. 512; Bob Otnes, "Keuffel & Esser–1880 to 1899," *Journal of the Oughtred Society* 10 (2001), pp. 18–28.

18 For further information about slide rules at fairs, see: Peggy A. Kidwell and Amy Ackerberg-Hastings, "Slide Rules on Display in the United States, 1840–2010", in Silke Ackermann, Richard L. Kremer, Mara Miniati (eds.), *Scientific Instruments on Display*, Leiden, 2014, pp. 159–172.

represented by holes punched in cards. Several fields of data could be entered on one card. After the data from one set of cards were counted, the cards could be sorted to count other characteristics of the population. Hollerith's tabulating system was designed for compiling statistics for the upcoming 1890 U.S. Census. As the Census occurs only every ten years, Hollerith not only sought the good will of international jurors but the attention of foreign governments. He won a gold medal and, in the ensuing years, some foreign business. The machine he exhibited was collected by the Conservatoire National des Arts et Métiers in Paris. However, his invention did not become truly profitable until it was improved and reworked into a tool of business and permanent government bureaucracies. Machines built from Hollerith's design were sold most successfully from the 1920s by a company which had come to have the name IBM.[19]

In 1893, the city of Chicago hosted a slightly belated celebration of the four hundredth anniversary of Columbus's arrival in this hemisphere.[20] At this fair, the German government aimed to demonstrate that its country not only boasted many eminent mathematicians, scientists, and educators, but also that it was at the very forefront of mathematical instrument making for these disciplines. German groups prepared several exhibits with their own particular guidebooks. Walther Dyck wrote one for the mathematics section of the German Universities exhibit.[21] Dyck, a student of Felix Klein, had already prepared an exhibition of mathematical apparatus for the 1893 meeting of the Deutsche Mathematiker-Vereinigung in Munich, with an extensive catalog of the objects shown.[22] When Klein was recruited to serve as a

19 Geoffrey D. Austrian, *Hermann Hollerith: Forgotten Pioneer of Information Processing*, New York, 1982, pp. 47–49. Hollerith also was asked to exhibit at the 1893 Columbian Exhibition in Chicago but declined, preferring to have his machines shown in the Census Department exhibit. They were accorded a bronze medal. See Ibid., pp. 100–101.

20 The importance of this fair to mathematics is discussed in: Karen Hunger Parshall and David E. Rowe, "Embedded in the Culture: Mathematics at the World's Columbian Exposition of 1893", *Mathematical Intelligencer* 15 (1993), pp. 40–45; Parshall and Rowe, *The Emergence of the American Mathematical Research Community 1876–1900: J. J. Sylvester, Felix Klein, and E. H. Moore*, Providence, R. I., 1994, esp. pp. 295–361; Ulf Hashagen, *Walther von Dyck (1856–1934): Mathematik, Technik und Wissenschaftsorganisation an der TH München*, Stuttgart, 2003.

21 Walther Dyck, *Special-Katalog der Mathematischen Ausstellung (Gruppe X der Universitäs-Ausstellung)*, Berlin, 1893.

22 W. Dyck, *Katalog mathematischer und mathematisch-physikalischer Modelle, Apparate und Instrumente*, Munich, 1893. This exhibit was planned for 1892, but an outbreak of cholera led to postponement of the event.

representative of German mathematics at the Chicago fair, he persuaded Dyck to arrange another exhibition, assisted by Oskar Bolza and Heinrich Maschke, two University of Chicago mathematicians who had studied under Klein at Göttingen. This thoroughgoing display included a large bust of Carl Friedrich Gauss; portraits of Carl Jacobi, Johann Peter Gustav Lejeune Dirichlet, and Bernhard Riemann; complete sets of seven German periodicals devoted in some way to mathematics; about 500 textbooks; and a copy of every doctoral dissertation on a mathematical subject written at a German university since 1850. There also were mathematical instruments, including a large sampling of the geometric models published by Ludwig Brill of Darmstadt *Pl. 6*).

The models made by Brill and his successor, Martin Schilling, are significant for several reasons. First, like the models for descriptive geometry designed by the Frenchmen Gaspard Monge and Théodore Olivier in the early nineteenth century, they were designed by mathematicians. Secondly, they were associated with the educational ideals of German universities, particularly those of mathematician Klein. Schools that purchased these models—in Great Britain, in Italy, in Japan, and in the United States—asserted their admiration of German methods. Thirdly, Brill models were purchased by many American mathematics departments for use in teaching relatively advanced topics. These were models for mathematicians.

Along with objects and publications, the German ministry of culture sent the eminent mathematician Felix Klein to the fair. He spent a certain amount of time expounding on the apparatus on display. He also presided over an international congress of mathematicians held on the fairgrounds. The mathematicians who actually attended the conference were primarily from the midwestern United States, but the program included papers contributed by Spanish, French, Czech, and, of course, German mathematicians. After the Congress, Klein gave a series of lectures at Northwestern University, and then toured the eastern United States.[23]

Klein and the other organizers of the Chicago exhibit proved apt exponents of German mathematical apparatus. In the summer of 1893 Seth Low, the President of Columbia College, visited the College's exhibit in Chicago. He was greatly taken with the nearby German exhibit and arranged at once to upgrade Columbia's display.[24] At the same time he purchased a collection of German

23 Parshall and Rowe, *The Emergence...*, op. cit. (n. 20), pp. 304–330.

24 Seth Low to the Trustees of Columbia College, December 4, 1893: Columbia University, New York, Rare Books and Manuscript Library, Columbia College Collection.

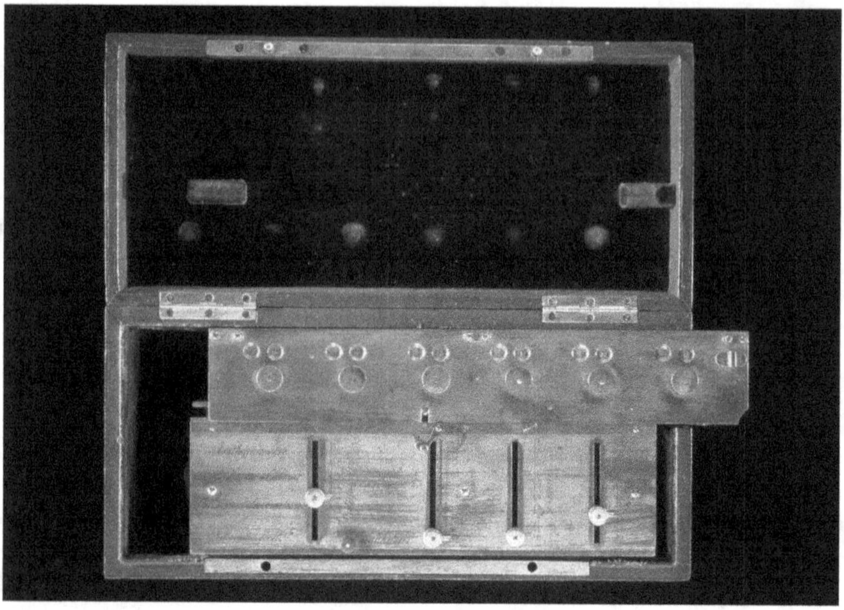

FIGURE 5.6 *Early Thomas arithmometer, exhibited in an historical exhibit at the Exposition universelle in Paris in 1900.*
SMITHSONIAN IMAGE 89-13225.

geometric models for the mathematics department.[25] Other schools, including Wesleyan University, the University of Illinois, Cornell, and Yale, also soon bought extensive collections of Brill models.[26]

Finally, if fairs could be used to promote mathematics, they also could describe the history of technology. In Paris in 1900 a retrospective exhibit of mathematical instruments included an adding machine of Blaise Pascal from 1652, borrowed from the Conservatoire National des Arts et Métiers. Also shown, on loan from the firm of Payen, was an 1820 model of a Thomas calculating machine (*Fig. 5.6*).[27] In St. Louis in 1904, several American makers of

25 Columbia College in the City of New York, *Fifth Annual Report of President Low to the Trustees*, October 1, 1894, p. 129.

26 A few institutions, such as Johns Hopkins, Clark University and Mt. Holyoke College, had them already. P. A. Kidwell, "American Mathematics Viewed Objectively: The Case of Geometric Models", in Ronald Calinger (ed.), *Vita Mathematica: Historical Research and Integration with Teaching*, Washington, D.C., 1996, pp. 197–207: 203–204.

27 Exposition Universelle Internationale de 1900, *Musée Rétrospectif de la Classe 13 Instruments de Précision. Rapport de Comité d'Installation*, Paris, 1900.

calculating devices exhibited both early and recent versions of their machines. For example, the National Cash Register Company showed a model of a cash register built by the Ritty brothers of Dayton, Ohio, that had been the starting point of their company. It is not clear whether the instrument shown in *Pl. 7* is the original model or a replica built for the fair.[28] The model did not change hands until decades later, when NCR gave it to the Smithsonian. However, NCR did also show its latest cash registers, as well as having them widely used on the fairgrounds.

To summarize, the late nineteenth and early twentieth centuries were a time of rapid transformation in mathematical instruments. The era also saw expanding international trade in devices from calculating machines to cash registers to drawing instruments to models of mathematical surfaces. World's fairs offered a point of exchange for these instruments, for knowledge of them, and for historical awareness. Indeed, instruments shown at fairs adorn collections of historic instruments to this day.

---

28 Kidwell and Ceruzzi, op. cit. (n. 1), pp. 40–41.

CHAPTER 6

Connections between the Instrument-making Trades in Great Britain and Ireland and the North American Continent

Gloria Clifton

There was trade in scientific instruments from England to North America from at least the early eighteenth century onwards and by the end of that century the leading makers were producing catalogues from which instruments could be obtained by mail order.[1] Even the government of the United States in the late eighteenth century obtained instruments from British suppliers, such as the hydrometers for measuring the alcoholic content of liquids which it acquired from John Dicas of Liverpool in 1790.[2] But alongside this long-distance trade there was also some relocation of craftsmen and a number of other means by which businesses established a physical presence on the other side of the Atlantic. Instrument makers moved from Great Britain and Ireland to the North American continent during the eighteenth and nineteenth centuries for a wide variety of reasons, ranging from transportation as a punishment for crime to bankruptcy or the hope of greater opportunities. From the mid-nineteenth century some British firms set up branch businesses in North America, both in the United States and in Canada, reflecting the desire of an expanding enterprise to increase sales in new markets. By the end of the nineteenth century there was also traffic in the opposite direction, with a few American instrument firms establishing offices in London. Another method firms used to try to enter new markets was to create an agency. This article will look at some of the sources which can be used to track these relocations and relationships from the eighteenth century to the early 1900s and try to assess whether there were any clear trends, or whether there was simply a series of random individual motivations.

Silvio Bedini and Deborah Warner have already tracked quite a number of individual instrument makers who went from Great Britain and Ireland to the American colonies or later the United States, and the reasons behind some

---

1 A. D. Morrison-Low, *Making Scientific Instruments in the Industrial Revolution*, Aldershot, 2007, pp. 164–166.

2 Ibid., p. 69.

 | DOI 10.1163/9789004324930_007

of these moves.[3] Those traced were generally individuals who either advertised that they had come from Great Britain or were listed as having been born there, or in Ireland, in American sources. There are very few documents which explain why a particular instrument maker emigrated, except in the case of those who, like Anthony Lamb in the early eighteenth century, fell foul of the law and had emigration forced upon them in the form of transportation to a penal colony. But sometimes it is possible to make some suggestions by linking people with the same name who appear in different kinds of sources, such as trade directories, London guild records and advertisements.

London, by the late eighteenth century and during the nineteenth century, was one of the leading centres in the world for the production of scientific instruments, but there were many firms and stiff competition. North America, which was expanding rapidly, probably seemed a more inviting prospect to an ambitious young man keen to set up his own business. Deborah Warner has shown that a number of these immigrants were indeed young men who had recently completed their apprenticeships.[4] There is also evidence that a number of those who arrived in North America from about 1830 onwards were interested in self-improvement, since they had attended classes at the London Mechanics' Institution, which had been opened in 1824 to provide evening classes in technical subjects and higher level studies for those who were at work during the day. Laurence C. Francis, who set up as an instrument maker in Philadelphia in the 1830s, had been apprenticed in 1824 to Peter William Piggott of Penton Street, Walworth, London, a drawing instrument maker and a freeman of the Merchant Taylors' Company, one of the ancient London guilds. During his apprenticeship he attended the London Mechanics' Institution from March 1827 to December 1829. There is no further record of him in London, and as the name is not very common, it is reasonable to assume that he is the same person as the one who appeared in Philadelphia a few years later.[5] In making these kinds of connections I have adopted the practice of historical demographers, who assume a link if three pieces of information referring to the same name are consistent, allowing for minor spelling variations, and there is no

3 Silvio A. Bedini, *Early American Scientific Instruments and their Makers*, Washington, D.C., 1964; Deborah J. Warner, "British Instrument Makers in the U.S.", *Rittenhouse: Journal of the American Scientific Enterprise* 6 (1992), pp. 25–31.

4 Ibid., p. 25.

5 Merchant Taylors' Company, "Apprentice Bindings", 1762–1933: London, Guildhall Library, Microfilm 320; London Mechanics' Institution, "Registers", 1824–1851: London, Birkbeck College (I am very grateful to Anita McConnell for sharing her notes from this source); Warner, op. cit. (n. 3), p. 28.

contradictory evidence. In this case, the only Laurence C. Francis who was the right age to be apprenticed in 1824 was Laurence Crokatt Francis, son of Richard, who was baptised in London in 1810. This agrees with the apprenticeship record, where the father's name is also given as Richard. Later, the age of the Laurence C. Francis recorded in United States censuses is consistent with this baptism record.[6]

Another example of a self-improver was probably James Green, who attended the London Mechanics' Institution from June 1830 to March 1831. Deborah Warner records a James Green visiting the leading instrument makers in London and Paris in 1831; encouraged by their assessment of the demand for instruments in the United States, he then emigrated with his brother and set up as an optician in Baltimore. Similarly, James Prentice who was in business in New York in the mid-nineteenth century as a drawing instrument specialist was almost certainly the same person as the man of that name who enrolled at the London Mechanics' Institute in 1828 and gave his occupation as drawing instrument maker.[7] In some cases the nature of their employment in England might have suggested to young instrument makers the potential opportunities in North America. For example, Benjamin Stancliffe had worked for Edward Troughton of London on the manufacture of instruments for the U.S. Coast Survey, before travelling to Philadelphia in 1816 and setting up his own business.[8]

It was also young men just out of their apprenticeship and looking for opportunities who were attracted to Canada in the mid-nineteenth-century. These were individual emigrants, such as Charles Potter, who came from a family of London instrument makers and opticians, and arrived in Toronto in 1853, when he was twenty-one years old. He soon established a partnership with William Hearn, who already had a watch-making and jewellery business in the city.[9] After Hearn's death he carried on the business alone, initially at 85 King Street West and later at 20 King Street East. In 1868 he was a founder member of the Toronto Astronomical Club, later the Royal Astronomical Society of Canada. Not all arrivals were young men, others sought opportunities in Canada in middle life, such as George Francis Berry, a compass adjuster in West

6 "St Mary, Islington, Register of baptisms", January 1801–December 1812: London, London Metropolitan Archive, P83/MRY1/1170, fol. 38.

7 London Mechanics' Institution, op. cit. (n. 5); Warner, op. cit. (n. 3), p. 28.

8 Ibid.

9 Julian A. Smith, "Charles Potter, Optician and Instrument Maker", *Journal of the Royal Astronomical Society of Canada* 87 (1993), pp. 14–33: 15–20.

Hartlepool, on the northeast coast of England, who was about fifty when he emigrated in 1912.[10]

Some of the instrument makers who moved to North America can be seen as serial migrants, in that they changed countries more than once in their careers. Timothy Claxton was one such example. He began his career in London, England, spent three years in Russia, in St. Petersburg, in the early 1820s, then moved to the United States, where he was in business in Boston selling philosophical instruments for thirteen years, before returning to London and re-establishing himself there.[11] Another maker who changed countries more than once was Joshua Ronchetti, who came to England from Italy in about 1805 and after a short period in the port of Liverpool settled in near-by Manchester, where he ran a successful instrument business for many years. But in 1854, when he was already in his sixties, he decided to move to New York, where he opened a business making thermometers, saccharometers, and other glass apparatus.[12] James William Merfield had a business in London in the 1840s as an optician and optical turner. By about 1850 he was in the United States, but in 1853 he moved to Melbourne in Australia and eventually became a telegraph contractor there.[13] Perhaps in these cases the motivation lay in their individual characters, in that they may have been people who easily became bored and began to look for a new challenge.

Sometimes, instrument makers who had crossed the Atlantic retained links with firms for whom they had worked previously and advertised that they could supply goods by the leading British makers. In a sense it was a kind of informal agency, but there is no evidence that the initiative came from the British firms concerned and so it was more likely a case of new arrivals trying to make their mark by offering a unique service. According to Deborah Warner, James Prentice, already mentioned as a drawing instrument maker in New York, advertised that he was "from Elliott's London" and that he had

10 T. N. Clarke, A. D. Morrison-Low, A. D. C. Simpson, *Brass & Glass: Scientific Instrument Making Workshops in Scotland as illustrated by instruments from the Arthur Frank Collection at the Royal Museum of Scotland*, Edinburgh, 1989, p. 157.

11 Timothy Claxton, *Memoir of a Mechanic. Being a sketch of the life of Timothy Claxton, written by himself. Together with miscellaneous papers*, Boston, MA, 1839.

12 Jenny Wetton, "Scientific Instrument Making in Manchester 1790–1870", *Manchester Memoirs* 130 (1990–91), pp. 37–68: 49–50.

13 Information from the late Professor Bert Bolton of Melbourne University. "Census of England and Wales" 1891: Kew, London, The National Archives, RG12/4103, fol. 118; "Death certificate", 16 April 1901: Kew, London, The National Archives (information from V. Majumdar).

"English Field Instruments constantly on hand".[14] Given his trade and the fact that he was offering surveying instruments for sale, he almost certainly came from the firm established by William Elliott, later Elliott Brothers of the Strand, London, which specialised in drawing and surveying instruments in its early years.[15] Another example was Charles Potter in Toronto, who had worked for the Dollond family of opticians in London and acted as their agent in Canada, a fact which featured in his advertising and that of Hearn and Potter.[16]

By the late nineteenth century the pattern of links between instrument making in the Great Britain, Ireland, and North America was altering, reflecting the changes in business organisation. There were fewer examples of the individuals who travelled across the Atlantic, then setting up their own instrument-making businesses. Rather, the larger firms were either setting up branches overseas or establishing agencies. Often, this was part of a pattern of establishing a presence for the firm across the globe, in which North America was only one of a series of locations. Such links were actively promoted by the British and Irish firms concerned, in contrast to the earlier situation, which was marked by individual initiative.

By the late nineteenth and early twentieth century, printed catalogues, and occasionally advertisements, are an important source for tracing the geographical expansion of businesses. Catalogues produced by some of the larger British instrument making firms often list their branches in the United States or Canada, or announce that they had appointed local firms as agents for their products. The microscope specialists, Richard and Joseph Beck, of Cornhill, London, issued catalogues in about 1880 as R. and J. Beck, manufacturing opticians, London and 1016 Chestnut Street Philadelphia, and in 1881 and 1882, just as R. and J. Beck, Philadelphia. This suggests that they were trading, or at least advertising, under their own name for a time in the United States, but their 1885 catalogue refers to R. and J. Beck, "Sole American Agents W. H. Walmsley and Co., 1016 Chestnut Street, Philadelphia, U.S.A.". So by this date they had clearly decided it was more advantageous to advertise through an American agent, rather than trying to market in their own name. They were still trading through American agents in 1890, but by then their representatives were Williams, Brown and Earle, corner of Tenth and Chestnut Streets,

14 Warner, op. cit. (n. 3) p. 28.

15 Gloria Clifton, *Directory of British Scientific Instrument Makers 1550–1851*, London, 1995, p. 95.

16 Smith, op. cit. (n. 9), pp. 15 and 17–18.

Philadelphia.[17] Similarly, manufacturing opticians, J. P. Cutts, Sutton and Son, based in Sheffield in the north of England, advertised in trade directories in the 1850s that they were also at 248 Pearl Street, New York.[18] I have not been able to trace an American instrument maker at that address, so unless further evidence is found, it looks as if this was a branch business. Similarly, the electrical instrument specialist, Robert William Paul, advertised in his catalogue of 1911 that as well as showrooms at 33 Leicester Square, London, and a factory at New Southgate, North London, he had premises at No. 1 East 42nd Street, New York.[19]

Other British and Irish firms who were active in North America always seem to have worked through agents. The London microscope maker Henry Crouch was advertising in his catalogues in the 1870s that his U.S. agents were James W. Queen and Company of Philadelphia.[20] There is also evidence that other American firms were agents for English products. The National Museum of American History has a Cambridge Scientific Instrument Company catalogue with a sticker for Arthur H. Thomas Company, 12th and Walnut Street, Philadelphia (*Fig. 6.1*). A later catalogue of the latter listed forty-five European firms whose instruments they imported.[21]

A number of the leading London instrument-making firms had outlets in Canada. Watts and Son Ltd., Surveying Instrument Makers, advertised in their catalogue of 1915 that they also had premises in Toronto and Ottawa (*Fig. 6.2*). Indeed it is clear that some major British firms thought it worth advertising in their catalogues that they were suppliers to various Canadian institutions, suggesting well-organised export arrangements even if they did not actually have premises or formal agencies in North America. C. F. Casella's catalogue announced that the firm was supplier to the Canadian Government (*Fig. 6.3*), while Negretti and Zambra thought it worth pointing out that they provided instruments for Toronto Observatory and had won prize medals at Philadelphia in 1876.[22]

17 R. G. W. Anderson, J. Burnett and B. Gee, *Handlist of Scientific Instrument-Makers' Trade Catalogues 1600–1914*, Edinburgh, 1990, p. 9.

18 Kelly & Co., *Post Office Directory of Sheffield*, London, 1854. I am grateful to A. D. Morrison-Low for this reference.

19 Anderson, Burnett and Gee, op. cit. (n. 17), p. 63.

20 Ibid., p. 21.

21 Ibid., p. 84.

22 Negretti & Zambra, *Encyclopaedic Illustrated and Descriptive Reference Catalogue of Optical, Mathematical, Philosophical, Photographic, and Standard Meteorological Instruments, Manufactured and Sold by them*, London, c. 1879, title pages.

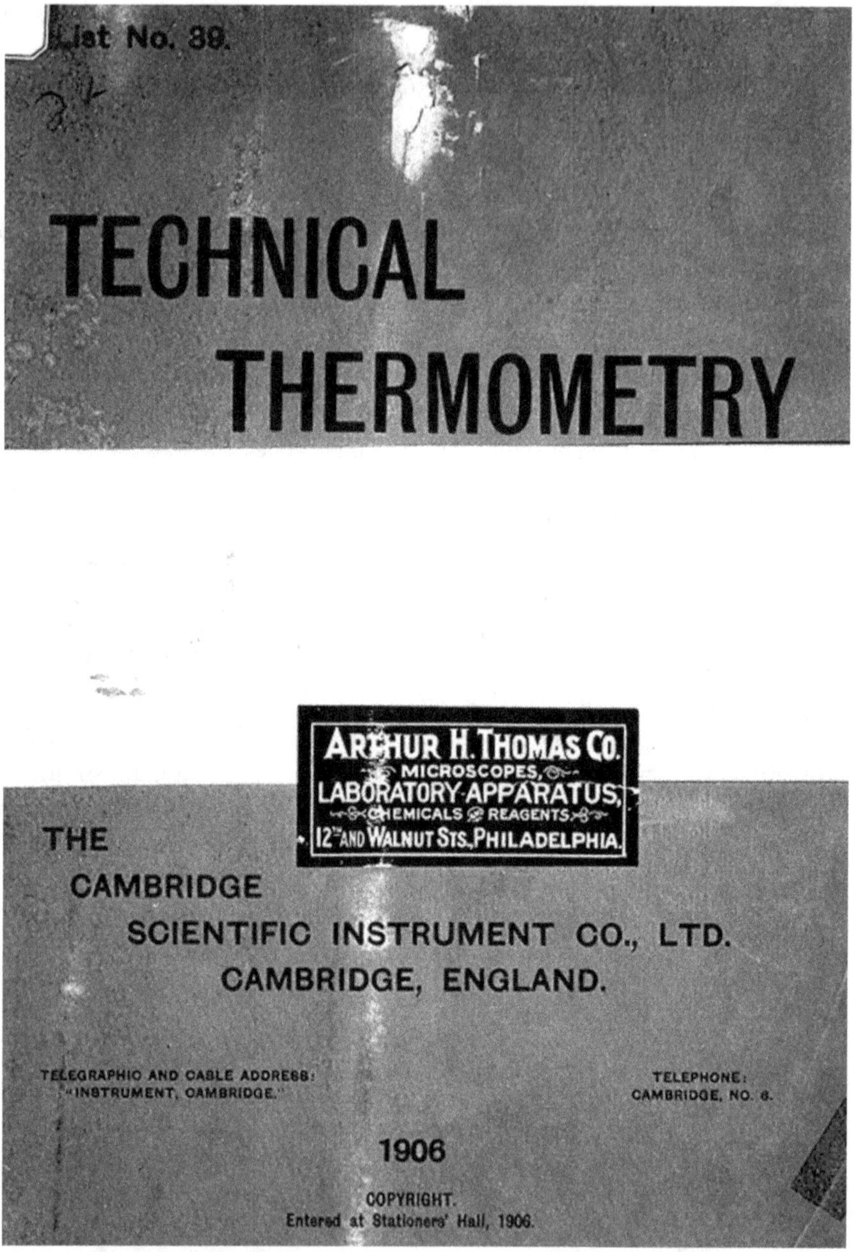

FIGURE 6.1 *The Cambridge Scientific Instrument Co. Ltd.*, Technical Thermometry, *Cambridge, UK, 1906.*
REPRODUCED BY PERMISSION OF THE SMITHSONIAN INSTITUTION, WASHINGTON, D.C.

CATALOGUE A137 *ESTABLISHED 1857*

ALL PREVIOUS LISTS CANCELLED

E. R. WATTS & SON LTD.

104 Victoria Street, Westminster

LONDON

ENGLAND

Canada: Instruments Ltd., Ottawa and Toronto

Belgium: Instruments de Précision Anglais, Brussels

Works: Camberwell, London

ERW&S

TRADE MARK

Manufacturers of Surveying and other Instruments of Precision

Codes: A B C, 5th and 6th Editions, and Bentley's

Telegrams and Cables: "COLLIMATOR," LONDON

TELEPHONE: HEAD OFFICE & SHOWROOMS, VICTORIA 0760

" Works - - - Rodney, 2301 (2 lines)

Printed by John Bellows, Gloucester

FIGURE 6.2 *E. R. Watts & Son Ltd.*, Manufacturers of Surveying and Other Instruments of Precision, *London, 1915.*

REPRODUCED BY PERMISSION OF THE NATIONAL MARITIME MUSEUM, GREENWICH, LONDON.

SURVEYING AND DRAWING INSTRUMENTS

C. F. CASELLA & Co., Ltd.

MAKERS OF

SURVEYING, METEOROLOGICAL & OTHER SCIENTIFIC INSTRUMENTS

TO

The Admiralty, Ordnance, Office of Works and other Home Departments, and to the Indian, Canadian and all Foreign Governments.

---

11 to 15, Rochester Row, Victoria Street, London, S.W.

1911

Established 1810.

FIGURE 6.3 *C. F. Casella & Co. Ltd.*, Surveying and Drawing Instruments, *London, 1911*. REPRODUCED BY PERMISSION OF THE NATIONAL MARITIME MUSEUM, GREENWICH, LONDON.

In some cases these kinds of references were clearly part of a marketing campaign aiming to depict the firm as a major international player. Hilger's catalogue of 1913 listed institutions all over the world which had bought their instruments and had an insertion listing their charges in the United States (*Fig. 6.4*). Sir Howard Grubb, based at Rathmines, Dublin, Ireland, a leading maker of large observatory instruments in the late nineteenth and early twentieth century, advertised in his 1888 catalogue that he was "Astronomical Instrument Maker and Contractor to the British, Colonial and Foreign Governments," and listed observatories and other institutions around the world that the firm had supplied (*Fig. 6.5*).[23] They also advertised their products through international fairs, gaining a Grand Prize in 1904 at the St Louis Exhibition.[24]

However, by the early twentieth century the instrument selling traffic went in both directions across the Atlantic. The Bausch and Lomb Optical Company of Rochester, New York, had agents in England by 1900, when their catalogue advertised A. E. Staley and Company of 35 Aldermanbury as London suppliers of their products.[25] The Weston Electrical Instrument Company of Newark, New Jersey, issued a catalogue in about 1910 which stated that they had a London office and laboratory at Audrey House, Ely Place, Holborn.[26] John Bliss and Co. advertised their mechanical log in *Brown's Nautical Almanac*, naming James Imray and Son as their British agents (*Fig. 6.6*). In some cases it is clear that American instrument manufacturers were very actively promoting British markets for their products. Elmer Sperry came to England in person in 1911 to try to obtain orders for his new gyro compass, and by 1913 he had opened a London office at 57 Victoria Street to handle his European business. Trials of his gyro compass were conducted by the British Navy in 1913, which resulted in orders for the American firm. Towards the end of World War I, Sperry set up a factory at Shepherd's Bush in the western suburbs of London.[27] However, trade in instruments continued to go in both directions after World War I, and

23 Sir Howard Grubb, *An Illustrated catalogue of Astronomical Instruments, Observatories, etc., manufactured by Sir Howard Grubb*, Rathmines, Dublin, 1888, title page.

24 J. E. Burnett and A. D. Morrison-Low, *'Vulgar and mechanick'. The Scientific Instrument Trade in Ireland*, Edinburgh, 1989, pp. 94–109.

25 Anderson, Burnett and Gee, op. cit. (n. 17), p. 8.

26 Ibid., p. 90.

27 Anthony E. Fanning, *Steady As She Goes: A history of the Compass Department of the Admiralty*, London, 1986, pp. 179–180.

2 **WAVELENGTH SPECTROMETERS.**

OTHER EDUCATIONAL ESTABLISHMENTS—*continued.*

Albert Agricultural College, Dublin.
University College of S. Wales, Cardiff.
University College of N. Wales, Bangor.
University College, Aberystwyth.
Technische Hochschule, Danzig.
Faculté des Sciences, Nancy.
Faculté des Sciences, Dijon.
Académie de Medécine, Paris.
Faculté des Sciences, Rennes.
Faculté des Sciences, Lille.
Technical College, Delft.
Polytechnic School, Zurich.
Technical Institute, Bari, Italy.
Lab. de Investigaciones Fisicas, Madrid.
School of Pharmacy, Lisbon.
Technische Hochschule, Moscow.
Landwirtschaftlisches Institut, Moscow.
Académie des Sciences, St Petersburg.
Electro-Technical Institute, St Petersburg.
Comité Geologique, St Petersburg.
Polytechnic School, Warsaw.
Mount Wilson Observatory, California, U.S.A.
Dartmouth College, New Hampshire, U.S.A.
Girard College, U.S.A.
Bowdoin College, U.S.A.
Polytechnic, Rio de Janeiro.
Central College, Bangalore.
Presidency College, Calcutta.
Government College, Lahore.
Engineering College, Sibpur.
Medical College, Madras.
Muir Central College, Allahabad.
Royal Observatory, Cape of Good Hope.
Transvaal University College.
Government of New South Wales.

INDUSTRIAL LABORATORIES.

Royal Arsenal, Woolwich (3 *instruments*).
British Home Office Laboratory.
Borough of Southwark (London) Analytical Laboratory.
Messrs Curtis's & Harvey, Cliffe at Hoo.
Messrs Kynoch, Ltd.
Messrs Nobels' Explosives Co., Ltd., Stevenston.
The E.C. Powder Co., Ltd., Dartford.
Elton Paper Mills, Bury, Lancs.
The Cotton Powder Co., Ltd., Faversham.
The United Alkali Co., Widnes.
Messrs Brunner, Mond & Co., Northwich.
Jandus Arc Lamp Co., London.
The Eastman Kodak Co., Rochester, U.S.A.
The L. E. Knott Apparatus Co., Boston, U.S.A.
The National Electric Lamp Co., Ohio, U.S.A.
The Bureau of Standards, Washington, U.S.A.
The Bureau of Soils, Washington, U.S.A.
La Société de l'Air Liquide, Boulogne s. Seine.
Geological Survey Office, Pretoria.
Agricultural Chemist of the Punjab.
Ordnance Laboratory at Naini Tal.
Forest Department, Government of India.

PRIVATE LABORATORIES.

The late Lord Blythswood, C.B.
Thomas M. Campbell, Esq.
Señor J. Munröz del Castillo.
A. A. C. Campbell Swinton, Esq., M.Inst.C.E.
Professor A. Cotton, Paris.
The late Dr F. H. Dupré.
W. Foster Brown, Esq., Newport, Mon.
J. W. Gifford, Esq.
Rear-Admiral Chas. E. Gissing.
Rev. J. M. Gordon.
The late Sir William Huggins, F.R.S., etc.
Professor H. Jackson, F.R.S.
T. R. Merton, Esq.
H. F. Parshall, Esq., M.Inst.C.E.
Dr H. A. Roome.
C. H. Stearn, Esq.

ADAM HILGER, Ltd., 75a Camden Road, London, N.W.

FIGURE 6.4 *Adam Hilger Ltd.*, Adam Hilger Ltd. [General Catalogue], *September 1913.*
REPRODUCED BY PERMISSION OF THE SMITHSONIAN INSTITUTION, WASHINGTON, D.C.

LIST OF SOME OF THE PRINCIPAL LARGE ASTRONOMICAL INSTRUMENTS MADE BY THE FIRM SINCE THE FOUNDATION BY THOMAS GRUBB

24 inch Reflector for Daramona, West Meath, Ireland.

15 inch Twin Refractor with 15 inch Photographic Doublet, for the late Mr. Dunn, Maidenhead. (Now at Hepple Woodside, Northumberland).

13 inch Photographic Refractors with 10 inch guiders for the International Survey of the Heavens, for Greenwich, Oxford, Melbourne, Cork, Perth (Western Australia), Mexico, and Cape Town Observatories.

15 inch Refractor for Mississippi, U.S.A.

7 metre Solar Spectrograph for Pulkovo Observatory, Russia, with Coelostat and second and third mirrors.

Spectroheliographs for Coimbra (Portugal), and Manchester.

15 inch Coelostat for the Marine Observatory, Kobe, Japan.

10 inch Coudé Telescope for the University Observatory, Cambridge, England.

30 inch Heliostat for the Smithsonian Institute, Washington, U.S.A.

Revolving Drums, Domes, etc., for the Edinburgh, Madrid, Greenwich, and many other Observatories.

[45]

FIGURE 6.5 *Sir Howard Grubb,* An Illustrated Catalogue of Astronomical Instruments, Observatories, etc., manufactured by Sir Howard Grubb, *Rathmines, Dublin, 1888.* REPRODUCED BY PERMISSION OF THE NATIONAL MARITIME MUSEUM, GREENWICH, LONDON.

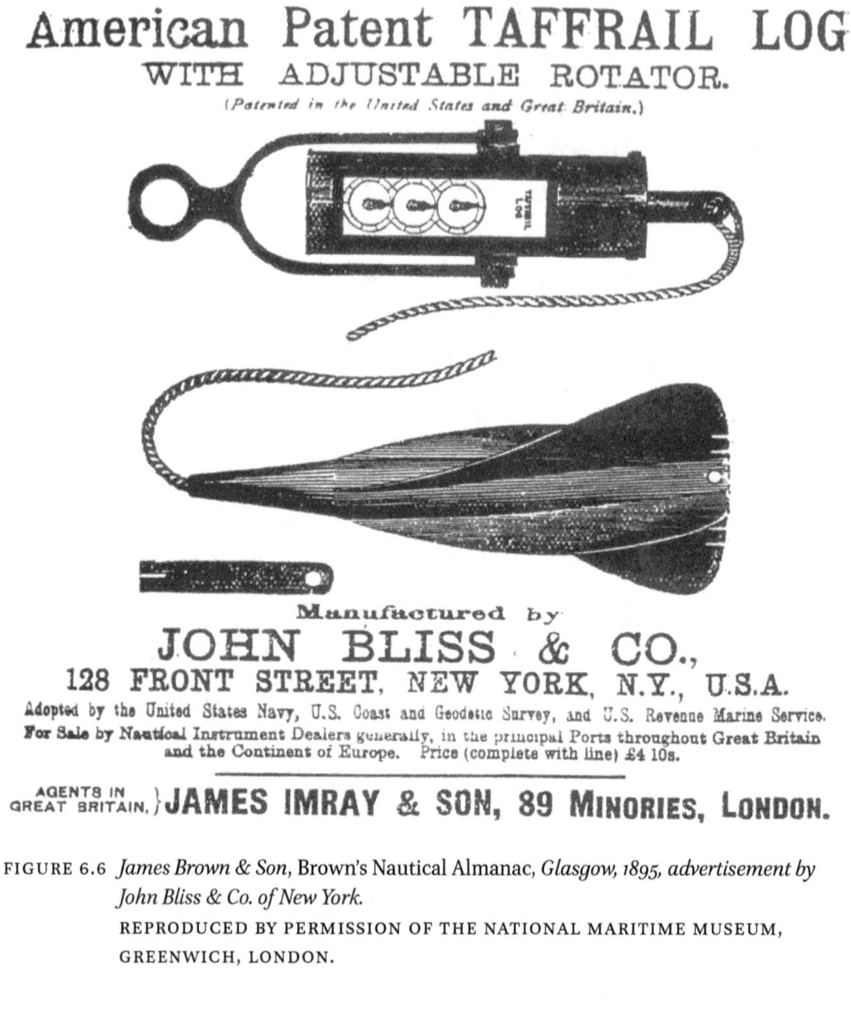

FIGURE 6.6 *James Brown & Son,* Brown's Nautical Almanac, *Glasgow, 1895, advertisement by John Bliss & Co. of New York.*
REPRODUCED BY PERMISSION OF THE NATIONAL MARITIME MUSEUM, GREENWICH, LONDON.

in the 1930s the nautical instrument suppliers, Kelvin, Bottomley and Baird, of Glasgow, had showrooms in Boston (*Fig. 6.7*).

It is not surprising to find that the trend in links between the scientific instrument trades in Great Britain and Ireland and those in North America mirrored the changes in the organisation of the industry on both sides of the Atlantic. The eighteenth and early nineteenth centuries saw individual immigrants to North America setting up their own enterprises to manufacture a range of scientific instruments, whereas there were fewer examples of individual initiative of this kind after about 1850. Rather, in the later nineteenth and

FIGURE 6.7 *Kelvin, Bottomley & Baird,* Navigational Instruments, *Glasgow, 1931.* REPRODUCED BY PERMISSION OF THE NATIONAL MUSEUMS OF SCOTLAND, EDINBURGH.

early twentieth centuries deliberate attempts were made by large instrument firms in Great Britain and Ireland to establish branches or agencies in North America to promote and sell their products. For the earlier period, establishing these links depends on tracing the individual histories of migrant workers through a range of sources, of which advertisements are just one element, with trade directories, censuses and records of births, baptisms, marriages, deaths, and apprenticeships being even more important. But the activities of the scientific instrument manufacturers of the late nineteenth and early twentieth centuries are most effectively traced through the catalogues, brochures, and advertisements they produced.

CHAPTER 7

European Pocket Sundials for Colonial Use in American Territories

*Sara J. Schechner**

Introduction

The first portable sundials brought to the Americas by European explorers and settlers were not made explicitly for use in those vast and wild lands, but were adapted for the purpose. What kinds were these, and when did types designed especially for use in America come to exist? Who needed or desired them? Where were they produced? What was their geographical range? To answer these questions, this paper will analyze archaeological evidence, household and business inventories, and most importantly, the very rare extant pocket sundials strongly linked to remote forts, tribal lands, battlefields, slave plantations, and colonial administrative seats. By means of close looking and critically thinking about this material culture, we can shed light on the relationship of time to imperialism and the transmission of cartographic and ethnographic knowledge during the colonial period.

* The author would like to thank Irene Castle McLaughlin, Curator of North American Ethnography at the Peabody Museum of Archaeology and Ethnology, Harvard University, for reading and commenting on a draft of this paper; Randall Brooks and David Pantalony for helping to locate sundials in Canadian museums; Dolph Druckman for permission to publish his family sundial; and Fei Carnes of the Center for Geographic Analysis, Harvard University, for help in drawing the maps. She would also like to thank the staff of the Adler Planetarium, Anne S. K. Brown Military Collection of Brown University Library, Canadian Museum of History, Collection of Historical Scientific Instruments at Harvard University, Fort Ticonderoga, Fortress of Louisbourg, John Carter Brown Library, Morristown National Historical Park, Neville Public Museum of Brown County, Wisconsin, New-York Historical Society, Peabody Museum of Archaeology and Ethnology at Harvard University, Preservation Virginia (Historic Jamestowne), Rhode Island Historical Society, and Stewart Museum.

 | DOI 10.1163/9789004324930_008

Seventeenth Century Stories

"Plantations in their beginnings have worke ynough, & find difficulties sufficient to settle a comfortable way of subsystence," Governor John Winthrop, Jr. wrote from Connecticut in 1668. No one had time or supplies to make pocket sundials when there were "buildings, fencings, clearing, & breaking up of ground to be attended, Orchards to be planted, High waies, bridges & fortifications to be made, & all thinges to doe, as in ye beginning of ye world".[1] A public sundial might be erected on a building, as shown by the Habitation of Quebec constructed by Samuel de Champlain in 1608 (*Fig. 7.1*) or by the crude slate dials unearthed at the site of Avalon, an English colony established in Newfoundland in 1621.[2] But pocket sundials? These were an indulgence brought to the New World by the colonists themselves in the first settlements and later by merchants in growing towns. The types transported to America's shores during the seventeenth century were those beloved in the Old World, and they fit their owners like a well-worn shoe. Three brief accounts of sundials in seventeenth century colonial America show this and establish a baseline from which sundial use diverged in the eighteenth century.

We begin with the most famous sundial: the one that brought Captain John Smith into contact with the Indian princess, Pocahontas, in 1607. It was an unusual, spherical, ivory compass sundial, and the dashing captain boasted that it had saved his life by fascinating his Indian captors in Virginia.[3] Smith's spherical sundial does not survive, but archaeological excavations of the James Fort site have yielded examples of rectangular, ivory diptych sundials made in Nuremberg (*Fig. 7.2*).[4] This more common form was apparently preferred

1 John Winthrop, Jr. to Henry Oldenburg, Secretary of the Royal Society, 12 November 1668, in A. Rupert Hall and Marie Boas Hall (eds.), *The Correspondence of Henry Oldenburg*, Madison WI, 1965–1986, 13 vols.; v. 5, p. 153.

2 Samuel de Champlain, *Les Voyages... de la Nouvelle France*, Paris, 1613, p. 187; and Sara J. Schechner, "Sundials of Newfoundland," unpublished paper presented at the annual meeting of the North American Sundial Society, Tenafly, New Jersey, August 19–22, 2004.

3 John Smith, *A True Relation of such occurrences and accidents of noate as hath hapned in Virginia*, London, 1608, p. B4r; J. Smith, *The Generall Historie of Virginia, New-England, and the Summer Isles*, London, 1624, sig. 47; S. J. Schechner, "The Adventures of Captain John Smith Among the Mathematical Practitioners: Cosmology, Mathematics, and Power at the Time of Jamestown," *Rittenhouse* 21, no. 2 (2009), pp. 126–144; S. J. Schechner, "The Adventures of Captain John Smith, Pocahontas, and a Sundial," *The Compendium* 14, no. 1 (2007), pp. 19–24.

4 Ivory diptych sundial, Hans Miller, Nuremberg, early seventeenth century, for latitude 53°, found in 2012 and discussed in Jamestown Rediscovery, Historic Jamestowne, Dig Updates, May 2012, http://historicjamestowne.org/may-2012-2/ (accessed 3 March 2016). Another

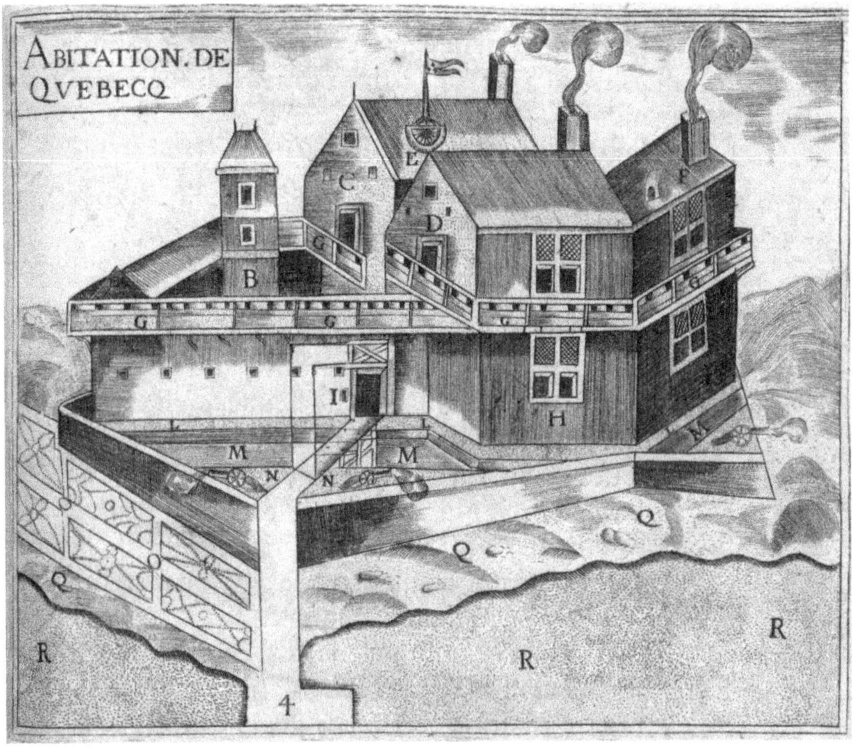

FIGURE 7.1 *Samuel de Champlain placed a public sundial beneath the flag on the roof of the fortified habitation he constructed at Quebec in 1608. Plate from Champlain,* Les Voyages . . . de la Nouvelle France, *Paris, 1613.*
HOUGHTON LIBRARY, HARVARD COLLEGE LIBRARY.

by the English gentlemen who accompanied Smith to Jamestown in 1607. The gnomonic projections on their sundials appear to be for latitudes more than 10° north of Jamestown (37° 12'), but accuracy in time finding was not critical in a wilderness fort. More important was the magnetic compass embedded in

ivory diptych sundial recovered since 1998, is described in Nicholas Luccketti and Beverly Straube, *1998 Interim Report on the APVA Excavations at Jamestown, Virginia*, Richmond, 1999, pp. 23–25; and Robert D. Hicks, "Instruments of Science and Technology in Early Virginia," *Rittenhouse* 21 (2007), pp. 65–81: 72–73. A brass compass box possibly part of an ivory sundial by Hans Ducher (1549–1632) or Hans Christoph Ducher (1584–1656), was excavated from a c. 1617–1624 context at the Governor's Land site adjoining Jamestown; see Alain Outlaw, *Governor's Land*, Charlottesville, VA, 1990, item 219. These publications include photos of the sundial fragments.

FIGURE 7.2 *In 2012, archaeologists in Jamestown, Virginia unearthed this ivory diptych from soil above a cellar dated to 1607–1610. It likely belonged to an English gentleman living in the early James Fort. The diptych is signed "Hans Miler" of Nuremberg and was designed for a latitude of 53°.*
COURTESY OF PRESERVATION VIRGINIA; HISTORIC JAMESTOWNE.

each diptych, which gave direction, and perhaps the fact that these ivory tools still had the capacity to impress other settlers of a lower status.

Our second example takes us north to New France. It is an oval ivory diptych unearthed from an Indian cemetery near the Jesuit House in Sillery, Quebec.[5] Aside from being ivory, it is simple and demure in design, and of a type commonly made in France in the mid-seventeenth century. What is its story? In 1637, Jesuit missionaries established a settlement in Sillery for indigenous peoples. Known as a 'reduction', the village was designed to Christianize and Europeanize its Montagnais and Algonquian residents through economic incentives like shelter, food, trade opportunities, and the use of a hospital staffed by nuns. But Jesuit efforts to turn hunters into farmers and impose

5 André E. Bouchard, "The Grand French Tradition in Gnomonics: The Presence in French America of a Scientific Tradition in the Art of Sundials," *The Compendium* 13, no. 2 (2006), pp. 3–11: 9.

European social values on Native Americans did not meet with success. By 1663 most Indians had left, and the village land was occupied by French settlers.[6] The discovery of the French oval ivory diptych in the Indian cemetery is fitting, but raises profound questions. Was it a gift to a native resident to encourage Western time discipline, or a timepiece lost by a missionary during a burial?

The last sundial story I wish to tell is that of Roger Williams, founder of the Providence Plantations that became Rhode Island. In Puritan New England, Williams advocated religious freedom, the separation of church and state, and the purchase, not confiscation, of land from Native Americans. These ideas—labeled as 'dangerous' by his Puritan neighbors—led to his banishment from the Massachusetts Bay Colony in 1636. Legend claims that Williams found his way to Providence with the help of his large brass compass sundial.[7] Made in England, the brass sundial was sturdy and economical. It was not extravagant in the way an ivory timepiece would have been in Puritan hands. The sundial also had a spooky *memento mori* in the center of a paper wind rose pasted inside the case lid. It featured a skull on top of a winged sand glass sitting on an urn (*Fig. 7.3*). This same motif is found in George Wither's *Collection of Emblemes, ancient and modern* (1635) with this motto: "*Live*, ever mindfull of thy dying; For, *Time* is always from thee flying".[8] How apt a sentiment this was for settlers who came to New England on an "errand into the wilderness"—their cause, the building of a model society by the elect on uncorrupted American soil.[9] The skull and winged sand glass served as a constant reminder not to waste time in vain pursuits, but to work hard to leave a positive mark on the world.

---

6 James P. Ronda, "The Sillery Experiment: A Jesuit-Indian Village in New France, 1637–1663," *American Indian Culture and Research Journal* 3, no. 1 (1979), pp. 1–18.

7 Compass sundial, English, owned by Roger Williams, c. 1630, brass, Rhode Island Historical Society, RHi X17 1147, Navigational Equipment, Museum 1902.3.1. It may date from the time of his later trips to England in 1643 and 1651–1654. Two similar compass dials signed I. R. are owned by the Collection of Historical Scientific Instruments, Harvard University, 7244 and 7253.

8 George Wither, *A Collection of Emblemes, ancient and moderne: Quickened with metracial illustrations, both morall and divine*, London, 1635, book 4, illustration 27.

9 "Errand into the wilderness" is a phrase taken from the title of a sermon preached in 1670 by Samuel Danforth, a Roxbury pastor, to delegates assembled for the election of officers to the General Court of Massachusetts. It expressed the notion that the Puritans had come on a religious mission to establish a pure reformed church in North America. See Samuel Danforth, *A Brief Recognition of New-Englands Errand into the Wilderness; Made in the Audience of the General Assembly of the Massachusetts Colony, at Boston in N. E. on the 11th of the third Moneth, 1670. Being the Day of Election there*, Cambridge, 1671; and Perry Miller, *Errand into the Wilderness*, Cambridge, MA, 1956.

FIGURE 7.3 *Roger Williams, the founder of Rhode Island, owned this English brass compass sundial, c. 1630. The image of a skull on top of a winged sand glass reminded him to spend his time wisely.*
COURTESY THE RHODE ISLAND HISTORICAL SOCIETY, RHI X17 1147.

The three examples show the role of pocket sundials as expressions of social status and religious sentiment, and as tools for missionary work, colonization, direction finding, and, of course, time finding. None of the principals in these three stories expected to travel very far once they arrived in their settlements. They had little need for a sundial adjustable for diverse latitudes. But with the growth of the American colonies came the diversification of purpose and justifications for the settlements. There arose new contracts with Europe and relationships among colonial societies. Up grew towns and ports where trade took place and people traveled. Therefore, by the eighteenth century we find portable sundials not only adapted to, but also designed for use in particular colonial marketplaces and locations. The remainder of this paper is concerned with these.

A Method of Latitude Analysis

One method for the discovery of sundials destined for use in colonial American territories is the examination of the latitudes for which they were made, particularly if those latitudes do not match the country of origin. Possible destinations are identified by means of the gnomonic projections on so-called 'particular' sundials designed for a fixed location, and by tables of latitude, known as gazetteers, associated with universal sundials. Such handy geographic directories were frequently engraved right on the surface of the sundial itself or on a metal disk mounted inside the lid of the sundial's case. Sometimes the directory was printed on a broadsheet, which was a sundial accessory.[10]

The variety of gazetteers has long interested me but been little studied. Very few are identical—even from the same workshop—and a comparison tells us something about production practices and the anticipated needs of owners.[11] In general, European makers listed cities in the regions where they made and sold the sundials, or they listed popular destinations throughout Europe where merchants might do business. A typical example is a silver sundial by Pierre le Maire II manufactured in Paris circa 1730–1760, and presently in the Collection of Historical Scientific Instruments at Harvard.[12] *Table 7.1* lists the cities on its gazetteer in the order that they appear on the underside of the sundial. The focus is on locations of great interest to French politics and trade. Out of twenty-six cities, only three were not under Bourbon rule sometime during the eighteenth century (i.e., Rome, Vienna, and London).

Pocket sundials for European use, such as the le Maire example, are common. Much more rare are sundials designed for use in colonial territories in North and South America. Some care, however, must be taken in the interpretation of their gazetteers.

10 Types of sundials with gazetteers include universal equatorial dials, diptychs, universal ring dials, Augsburg-type dials, Butterfield-type dials, and inclining dials.

11 The collections of the Adler Planetarium in Chicago (hereafter, AP), the Collection of Historical Scientific Instruments, Harvard University (hereafter, CHSI), and the National Maritime Museum (hereafter, NMM), Greenwich, and those of many other museums, offer numerous examples of sundials of the same type and quality produced in the same workshop with different geographical listings piece by piece. The gazetteers are transcribed and photographed in S. J. Schechner, *Sundials and Time-Finding Instruments of the Adler Planetarium* (forthcoming); Steven A. Lloyd, *Ivory Diptych Sundials, 1570–1750*, Cambridge MA, 1992; *Waywiser*, the online catalogue of the Collection of Historical Scientific Instruments, http://waywiser.rc.fas.harvard.edu/ (accessed 3 March 2015); and Hester Higton, *Sundials at Greenwich*, Oxford, 2002.

12 Butterfield-type sundial, Pierre le Maire II, Paris, c. 1730–1760, CHSI inv. 7028. The Roman numeral 'II' is not formally part of le Maire's name, but used by historians to distinguish him from his father, also a mathematical instrument maker.

TABLE 7.1 *Gazetteer on a silver Butterfield-type sundial by Pierre le Maire II, Paris, c. 1730–1760*

Underside of compass box	Left column	Right column
Londres 51	Paris 48. 51	Lion 45. 45
Bruxelles 50. 51	Brest 48. 23	Turin 44. 50
Lisle 50. 40	Rennes 48	Milan 45. 20
Calais 50. 57	Basle 47. 40	Pau 43. 12
Liege 50. 36	Strasbourg 48. 35	Bayonne 43. 30
Nantes 47. 13	Vienne 48. 24	Montpellier 43. 37
	Roüen 49. 27	Perpignan 42. 44
	Bordeaux 44. 50	Madrit 40. 26
	Poitiers 46. 34	Naples 41. 5
	Dijon 47. 20	Rome 41. 54

Sundials for a Trip Around the World

One class of pocket sundials that may include American place names is the group whose gazetteers list cities and countries from all over the globe. These sundials are of universal or semi-universal functionality, which enables their use for most of the northern hemisphere. The examples discussed here are preserved at Harvard, the National Maritime Museum at Greenwich, and the Adler Planetarium in Chicago.

The first is a silver inclining dial by Delure, Paris, circa 1695–1736. Its forty-three-city gazetteer starts off with Paris, Brest, Dublin, Vienna, Cracow, and other major European destinations, but before it reaches Rome, it has also recorded: *Candie 34 40* (Crete), *Alep 36 15* (Aleppo, Syria), *Le Caire 30* (Cairo), *Ispaham 36 14* (Isfahan, Iran), *Siracuze 34 4* (Syracuse, Italy), *Pequin 40* (Beijing), *Smirne 38 22* (Smyrna, today Izmir, Turkey), *Feéz 33* (Fez, Morocco), and *Goa 15* (in India).[13] The second example is a silver Augsburg-type sundial by Michael Butterfield, Paris, circa 1675–1715. Among thirty-three places on its geographical directory are the following surprises: *Agra 20. 30* (in India), *Cape de Bonne Esperance 34. 15* (Cape of Good Hope), and seven places spread out in the Ottoman Empire, such as *Constantinople 41. 6*, *Babilonne 33. 40* (Babylon,

13 Delure, Paris, ca. 1695–1736, CHSI inv. 7356.

Mesopotamia); *Ierusalem 31. 50* (Jerusalem); *Alexandrie 31. 11* (Alexandria, Egypt), *Algere 36. 38* (Algiers), *Malte 35. 53* (Malta), and *Candie 35. 19* (Crete).[14]

Although these two objects do not include any American locations, similar sundials do. A brass Augsburg-type sundial by Delure includes *Kebec 47* (Quebec) along with outliers *Macao 40* (Macau), *Pequin 40* (Beijing), and *Cap. de. B. Esperance 34 15'* (the Cape of Good Hope) among thirty-eight cities that are otherwise in Europe or around the Mediterranean.[15] A Claude Langlois, silvered-brass, Augsburg-type dial, dating from the second quarter of the eighteenth century, not only lists many Mediterranean ports—including the uncommon *Detroit de Gibraltar 35.30'* (Strait of Gibraltar)—but also records five New France and New Spain possessions: *Kebec 47* (Quebec), *St Domingue 17. 38'* (Saint Domingue, now Haiti), *Lima 12. 20', Mcxiquc 20* (Mcxico City), and *Quito 00. 18'* (in Ecuador).[16]

The arbitrariness of the geographical directories is evident in the foregoing examples, and this point is underscored by several universal sundials constructed at the workshop of Nicolas Bion between 1685 and 1715 and presently owned by the National Maritime Museum. One is a brass Augsburg-type dial that catalogues locales in this order: *Paris 49, Vienne 48, Dantzic 54* (Gdansk), *Varsovie 52* (Warsaw), *Naples 41, Madrid 40, Rig 57* (Riga), *Londres 52* (London), *Lima 12, Cadis 36* (Cadiz), *Lahauane 23* (Havana), *Berlin 52, Pequin 40 (Beijing), Siam 14* (Thailand), *Cracovie 50* (Cracow), *La Martinique 15, Ponticheri 15* (Pondicherry, now Puducherry), *Goa 15, St Domingue 18, Strasbourg 49, Amsterdam 52, Ausbourg 48* (Augsburg), *Petersbourg 60* (St. Petersburg), *Hambourg 54*, and *Copenhague 56*.[17] Two of Bion's universal ring dials likewise include cities stretching from Asia to South America. One of these—a large brass example at Greenwich—intersperses *Lima 12, Canada 55, Pekim 40* (Beijing), *Moskou 55* (Moscow) and *Siam 18* (Thailand) among over forty European cities.[18] A smaller brass dial adds *Cayenne 6* (in French Guiana), *Martinique 18, Panama 10, Canada 49, Quebec 48, St Dominque 20*

14 Augsburg-type dial, Michael Butterfield, Paris, ca. 1675–1715, NMM inv. AST0524; Higton, op. cit. (n. 11), pp. 164–165.

15 Augsburg-type dial, Delure, Paris, ca. 1700–1736, AP inv. L-3.

16 Augsburg-type dial, Langlois, Paris, second quarter of the eighteenth century, NMM inv. AST0518; Higton, op. cit. (n. 11), pp. 168–169.

17 Augsburg-type dial, Nicolas Bion, Paris, ca. 1685–1715, NMM inv. AST0285; Higton, op. cit. (n. 11), pp. 165–166.

18 Universal ring dial, Nicolas Bion, Paris, ca. 1685–1715, NMM inv. AST0311; Higton, op. cit. (n. 11), pp. 202–203.

(Saint Domingue), as well as *Bornes 3* (Borneo), *Camboie 10* (Cambodia), *Goa 15*, and *Macao 22*.[19]

There are several points to be made. 1: Rarely when places are south of the equator are they marked as such, and some of the sundials that include them cannot operate in the southern hemisphere because they lack the counterclockwise hour scale required for use with a south-pole-aligned gnomon. 2: Cities and countries are given equal weight in gazetteers, although a single latitude cannot adequately inform a user of the geographical range covered by an entire people or country. 3: The examples noted above are all French, because English and German makers in the seventeenth and eighteenth centuries rarely produced universal sundials that recorded sites all over the globe. 4: French workshops, however, did not confine the exotic destinations to places of particular French interest. Although Quebec, Saint Domingue, and Pondicherry were French colonial possessions, Goa and Macau were under Portuguese influence; Peru and Mexico were Spanish colonies; and Isfahan and Algiers were part of the Ottoman Empire. 5: New England colonies, it seems, were altogether snubbed unless they were in territories contested by France and England. I have yet to see a French sundial from the seventeenth or eighteenth century that includes Boston, New York, or Philadelphia. These cities did not appear on European-made sundials until about 1800, when they were printed on the paper gazetteers glued to cheap wooden diptychs manufactured in Bavaria for export.[20]

Only an owner engaged in one of the East or West India trading companies would find practical use in a timepiece that included not only northern Europe but also the tip of southern Africa, Asia, and the Americas. Even so, it is implausible that any single individual would be traveling between Moscow, Lima, Quebec, Agra, and Dublin. Moreover, too many of these trip-around-the-world sundials survive for global trade to be an adequate reason to explain their existence. So why do these instruments list such widespread cities? What would be the attraction for buyers? The reasons to own such a sundial may be similar to those for owning a world atlas: the pleasure might come from possessing the world from the comfort of one's library or from nonchalantly displaying one's worldly awareness to associates. These cosmopolitan directories

19 Universal ring dial, Nicolas Bion, Paris, ca. 1685–1715, NMM inv. AST0296; Higton, op. cit. (n. 11), p. 202.

20 See diptych examples by David Beringer, Nuremberg, 1777–1821, AP inv. N-16; Stockert, Fürth, Bavaria, ca. 1800–1820, AP inv. N-17; Negelein, Nuremberg, ca. 1825–1850, AP inv. N-15; Würsching, Fürth, Bavaria, ca. 1825–1850, AP inv. A-150; and J. G. Kleininger, Nuremberg, ca. 1825, AP inv. W-202 and W-71.

likely spoke to their owners' fascination in exotic locales and peoples, just as pictures of Native Americans and Ottoman Turks symbolically did a century earlier on ivory diptych sundials (*Pl. 8*).[21]

We must assume, therefore, that few trip-around-the-world sundials ever saw American shores. We now turn our attention to more specialized sundials that clearly did.

Sundials for Sugar Plantations in the Antilles

Knowledge of sundials made explicitly for use in the Americas comes from extremely rare, extant objects found in institutional and private collections. The first group to be discussed tells a story about French colonial life in the West Indies.

We begin with an object at the Adler Planetarium in Chicago. It is an elegant, octagonal, silver Butterfield-type sundial signed *Baradelle AParis* (*Fig. 7.4*).[22] It has only two hour scales—one for 20° and another for 25°—and they are divided into quarter-hour intervals and labeled in Arabic and Roman numerals respectively. Adjacent to the inner scale, the maker has inscribed a single geographic name: *S*ᵗ*. Domingue. 18*ᵈ*.* The folding gnomon is partly missing, but what remains proves that it was adjustable for latitude. It could have once had a cute bird whose beak pointed to the latitude scale. There is no gazetteer on the underside of the sundial, but the maker has richly engraved a flower and acanthus leaves on the bottom of the compass box and the gnomon spring. The sundial has no records associated with its early history, but we can learn a lot by close looking and critically thinking.

The Baradelles were a distinguished family of mathematical instrument makers in Paris. Nicolas-Jacques Baradelle (1701–after 1772) was the senior member. He apprenticed with N. la Butte (in 1717), Jacques le Maire (also in 1717), and Nicolas Bion (from 1719 to 1725), and lived and worked at Bion's address on the Quai de l'Horloge in Paris. By 1752, he was operating down the street under the sign *à l'Observatoire* in homage to his godfather, Jacques Cassini, director of the Paris Observatory between 1712 and 1756. Baradelle also had close connections at the Académie des Sciences, and employed members' data in some of his instruments. His grandsons included the instrument makers

21 Two ivory diptych sundials, Thomas Ducher, Nuremberg, 1620–1645, CHSI inv. 7579 and 7573.

22 Silver Butterfield-type sundial, Baradelle, Paris, mid-to-late eighteenth century, AP inv. W-67.

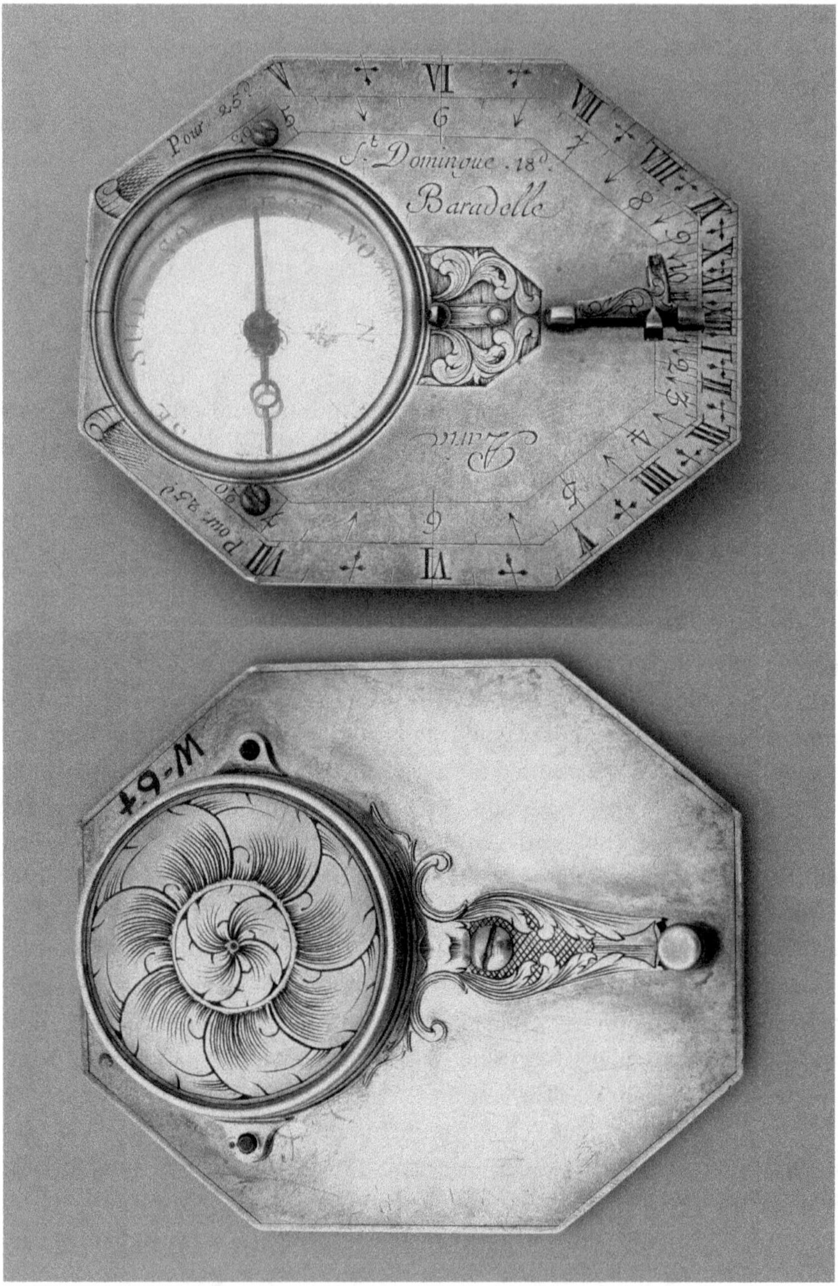

FIGURE 7.4 *Designed for use in the French colony of Saint Domingue (today Haiti), this silver Butterfield-type sundial was made in Paris by the Baradelle workshop in the mid-to-late eighteenth century.*
PHOTOGRAPHY BY STEVE PITKIN, PITKIN STUDIO. COURTESY OF THE ADLER PLANETARIUM, CHICAGO, ILLINOIS, W-67.

Nicolas-Alexandre Baradelle (ca. 1740–post 1791), Nicolas-Elois Baradelle (1749–1814), and Jean-Louis-Jacques Baradelle (1752–1794), all of whom were in the Founders' Company.[23] Any one of them could have made this sundial, but the best bet would be Nicolas-Jacques or Nicolas-Alexandre Baradelle, who were notable for their high-quality sundials. We therefore conclude that the Adler Planetarium's sundial dates from the mid-to-late eighteenth century.

The sundial's inscription, *S^{t}. Domingue. 18^{d}*, is another clue. Between 1659 and 1804, Saint Domingue was a significant French colony on the western side of the island of Hispaniola. It is now part of Haiti. The sundial's hour scale for 20° is a good match for Saint Domingue's first capital on Tortuga, an island just off the northwest coast of Hispaniola at latitude 20°, as well as its second capital, Cap-Français at 19° 45' on the mainland's northern shore (today Cap-Haïtien). In 1770, the colony shifted its government further south to Port-au-Prince (18° 32'), which may be the place referenced by the sundial's inscription, *S^{t}. Domingue. 18^{d}*. A brief look at the colony's history suggests the type of individual who might have owned and used this sundial.

In the first half of the seventeenth century, Tortuga traded hands many times between the Spanish, who claimed it, and French and English settlers who occupied it without permission. It was a notorious haven for renegades and runaways who used the island as a base for plaguing Spanish ships and a source of wild cattle, hogs, timber, and fresh water. From their practice of grilling meat on a *boucan* in the style of the island's indigenous Taíno people, these settlers came to be called buccaneers. Their first livelihood was trade in hides and cured meat, tobacco and crude sugar with anyone willing to buy outside of the law; their second and more notorious business was piracy.[24] In 1640, the French built Fort de Rocher in a harbor on Tortuga. In 1654, the Spanish recaptured the island, but by 1655 the English and French settlers were back, with the English claiming it as their own. In 1659, the English-appointed governor, a Frenchman, Jeremie Deschamps, claimed Tortuga for France.

In 1665, the French West India Company established a permanent settlement on the mainland of Hispaniola and set up a plantation-based economy in

---

23 Jean-Dominique Augarde, "La Fabrication des instruments scientifiques au XVIIIe siècle et la corporation des fondeurs", in Christine Blondel et al. (eds.), *Studies in the History of Scientific Instruments*, London, 1989, pp. 52–72; Denis Beaudouin, Paolo Brenni and Anthony J. Turner, *A Bio-bibliographical Dictionary of Precision Instrument-makers and Related Craftsmen in France, 1430–1960* (in preparation).

24 Kris E. Lane, *Pillaging the Empire: Piracy in the Americas, 1500–1750*, Armonk, NY, 1998, pp. 97–101; Jan Rogoziński, *A Brief History of the Caribbean from the Arawak and Carib to the Present*, rev. ed., New York, 1999, pp. 85–86, 92–94 and 99–100.

FIGURE 7.5 *Slaves at work on an indigo plantation in Saint-Domingue, 1667, under the watchful eye of their master. Plate from Jean Baptiste Du Tertre,* Histoire Générale des Antilles Habitées par les François, *Paris, 1667, and was reproduced three times in publications in the eighteenth century.*
JOHN CARTER BROWN LIBRARY.

which tobacco and indigo were raised. In 1697, Spain formally ceded the western third of Hispaniola to France in the Treaty of Ryswick. The capital of the French colony was moved to Cap-Français on the mainland's northern coast in 1711. It would become a city of wealth and sophistication in a colony nicknamed "the pearl of the Antilles".[25] Sugar cane and coffee joined indigo, cacao, and cotton as the colony's chief exports (*Fig. 7.5*). By the 1780s, Saint Domingue produced 40% of the sugar and 60% of the coffee consumed in Europe.[26] But this wealth was founded on an ugly fact: with large scale production came the need for increased labor—a need met by the colonists importing over 800,000

25 Médéric-Louis-Elie Moreau de Saint-Méry, *A Civilization that Perished: The Last Years of White Colonial Rule in Haiti*, translation and abridgement of *A Topographical, Physical, Civil, Political and Historical Description of the French Part of the Island of Santo Domingo, with General Observations on its Population, on the Character and Customs of Its Diverse Inhabitants, on its Climate, Culture, Production, Administration, Etc.*, by Ivor D. Spencer, Lanham, MD, 1985, pt. 3, pp. 93–158.

26 Rogoziński, op. cit. (n. 24), pp. 108, 116 and 118.

African slaves since 1680, many of whom died within a few years of reaching Saint Domingue. A slave rebellion in 1791 culminated in the establishment of the independent country of Haiti in 1804.[27]

With one in eight people in France earning a living touched by the country's trade with Saint Domingue, it is surprising that Saint Domingue is not found on more French sundials than it is.[28] The Baradelle sundial is therefore noteworthy. With hour scales for 20° and 25°, the silver sundial was perfectly suited to a wealthy, white, French merchant, colonial administrator, or plantation owner, living in or near Cap-Français, which was the chief port and *de facto* capital throughout the colonial period. The owner may also have been among the *affranchis* (freed black slaves) or *gens de couleur libres* (free people of mixed French and African race, such as mulattoes). Although the political rights of free blacks and people of color were limited, they could own property, shops, plantations, and slaves. Many were educated, and some were quite prosperous. Emulating the styles of their white French neighbors, wealthy free blacks and people of color, both men and women, would have found this silver sundial a desirable possession (*Fig. 7.6*). The biggest mystery of the sundial is not the wealthy status of possible owners, but the business that could take him or her as far north as 25°. That is the latitude of Nassau in the British colony of the Bahamas. Perhaps the owner was a French officer fighting the British in the Caribbean during the Seven Years' War (1756–1763).

The Baradelle sundial is a rare but not unique product made in Paris for use in the French West Indies. Recently, I learned of another portable sundial made explicitly for the latitude of 18° (*Fig. 7.7*). This window-sill sundial is made of brass and has a rectangular base. It is inscribed *Lenoir AParis* and *18^{d} latitude*.[29] Although it has a single hour scale marked in Roman numerals and divided into quarter hours, its gnomon is adjustable between 16° and 20° with subdivisions every 15 minutes. The magnetic compass has an eight-point wind rose labeled in French, a circle divided 0°-90°-0°-90° every 1°, and

---

27 Kona Shen, "History of Haiti, 1492–1805," timeline sponsored by the Brown University Department of Africana Studies, 2008, http://library.brown.edu/haitihistory/1sr.html (accessed 07 March 2016). M.-L.-E. Moreau de Saint-Méry, *Description topographique, physique, civile, politique et historique de la partie française de l'Isle Saint-Domingue*, Philadelphia, 1797–1798, 2 vols.; *Le code noir ou Edit du roy*, Paris, 1685, and later editions, issued by Louis XIV for regulation of slavery in Saint Domingue; Rogoziński, op. cit. (n. 24), pp. 127–142.

28 John Carter Brown Library, "Remember Haiti: Economy", www.brown.edu/Facilities/John_Carter_Brown_Library/remember_haiti/economy.php (accessed 07 March 2016).

29 Brass, rectangular portable sundial, 16.5 × 17.8 cm, Lenoir, Paris, ca. 1800–1825, private collection of Dolph Druckman, Baltimore, Maryland.

FIGURE 7.6 *The owner of a French silver sundial in Saint Domingue between 1750 and 1800 may have been a prosperous freed black slave or person of color, such as those depicted by Nicolas Ponce,* Recueil de vues des lieux principaux de la colonie françoise de Saint-Domingue, *Paris, 1791 (two details).*

a brake that lifts the diamond-shaped needle off its pivot when transported (a feature more typical of surveying compasses than sundials). The top and bottom of the sundial are plain with no adornments in open spaces. Whereas the silver Baradelle sundial has the air of luxury and refinement, this brass Lenoir instrument is very work-a-day, being more suited to a military encampment, a wharf, or a sugar mill. Etienne Lenoir (1744–1832) or his son Paul-Étienne Lenoir (1776–1827) probably supplied this utilitarian instrument in the first quarter of the nineteenth century to a business in Haiti. It is the only known sundial by Lenoir.[30]

Also known are two pocket sundials designed for use in Martinique, a French possession since 1658 and another center for the production of sugar.[31] One held by the National Maritime Museum in Greenwich is a finely-engraved, brass, Augsburg-type dial made by Claude Langlois, in the second quarter of

30 A. J. Turner, *From Pleasure and Profit to Science and Security: Etienne Lenoir and the Transformation of Precision Instrument-Making in France, 1760–1830*, Cambridge, UK, 1989.

31 Robert Cornevin, "Martinique", in *Encyclopaedia Britannica Online*, www.britannica.com/EBchecked/topic/367084/Martinique/54686/French-rule (accessed 9 March 2016).

FIGURE 7.7 *A sundial for Haiti made by the Lenoir workshop in Paris in the early nineteenth century.*
COLLECTION OF DOLPH DRUCKMAN, BALTIMORE, MARYLAND.

the eighteenth century. On its long list of European cities, there is an interloper: *La Martinique 14. 43'*.[32] The single colonial destination on the sundial suggests that its owner had business that took him from France to the sugar plantations in Martinique. Perhaps he was a merchant or soldier. A private collection holds the second example, which was made in 1764 by Jacques Canivet (fl. 1743–1774), the successor to Langlois as chief engineer to the Académie des Sciences in Paris. This is an octagonal silver sundial for exclusive use at the latitude of

32 Brass Augsburg-type dial, Langlois, Paris, second quarter of the eighteenth century, NMM inv. AST0260; Higton, op. cit. (n. 11), pp. 169–170.

14° 45', which is the location of Martinique.[33] Its possessor may have been a French officer or administrator stationed on the island after the Seven Years' War, or a prosperous plantation owner or merchant, who was likely white but may have been a freed black or person of color.

Sundials for the Administration of New Spain along the South Sea

The Harvard Collection of Historical Scientific Instruments owns a mid-eighteenth century sundial designed to be used along the Spanish frontier in colonial Chile.[34] Most likely made in Spain, this rare, brass, universal ring dial has its gnomonic projections flipped correctly for the southern hemisphere (*Pl. 9*). Both sides of the meridian ring are engraved with sixteen place names between the latitudes of 25° S and 41° S. Nearly all the locations are along the Pacific coastline (then known as the South Sea), but several are in the interior, including Santiago, Maule, and Chillán. The gazetteer also lists Mendoza and San Juan de la Frontera, which were at the time in Paraguay and today are in Argentina. Before 1776, all of the sundial's locations were under the jurisdiction of the Viceroyalty of Peru, and through it, the kingdom of Spain. Santiago was the capital city of Chile and the seat of a royal *audiencia* from 1609 until the country's independence (*Table 7.2* and *Fig. 7.8*).[35]

Chile was an agricultural colony, but had one of the largest standing armies in the Americas. It was heavily militarized in order to protect the natural resources from Spain's European enemies and to fend off the indigenous Araucanian people in the south, who fiercely resisted the Spanish settlers.[36] As we saw in the French West Indies, the population of Chile was divided along ethnic, racial, and class lines. At the top were European-born Spaniards (*peninsulares*) living in the colony as royal administrators and Chilean-born Spaniards (*criollos*), who were landowners. At the bottom were Native

33 Silver, octagonal, pocket sundial for Martinique, 14° 45', signed "Canivet à la Sphère à Paris / 1764," pictured in Mike Cowham, *A Dial in Your Poke*, Cambridge, UK, 2004, p. 68; Franck Marcelin, *Dictionnaire des fabricants français d'instruments de mesure du XVe au XIXe siècle*, Aix-en-Provence, 2004, s.v. "Canivet, Jacques."

34 Universal ring dial, Spanish, 1725–1775, CHSI inv. 7904.

35 Emanuel Bowen, *A New and Accurate Map of Chili, Terra Magellanica, Terra del Fuego &c. Laid down according to the latest Improvements and Regulated by Astron.l Observat.ns*, London, 1747: David Rumsey Historical Map Collection, 3733.053.

36 Rex A. Hudson (ed.), *Chile: A Country Study*, Washington, DC, 1994: "Conquest and Colonization, 1535–1810," http://countrystudies.us/chile/ (accessed 07 March 2016).

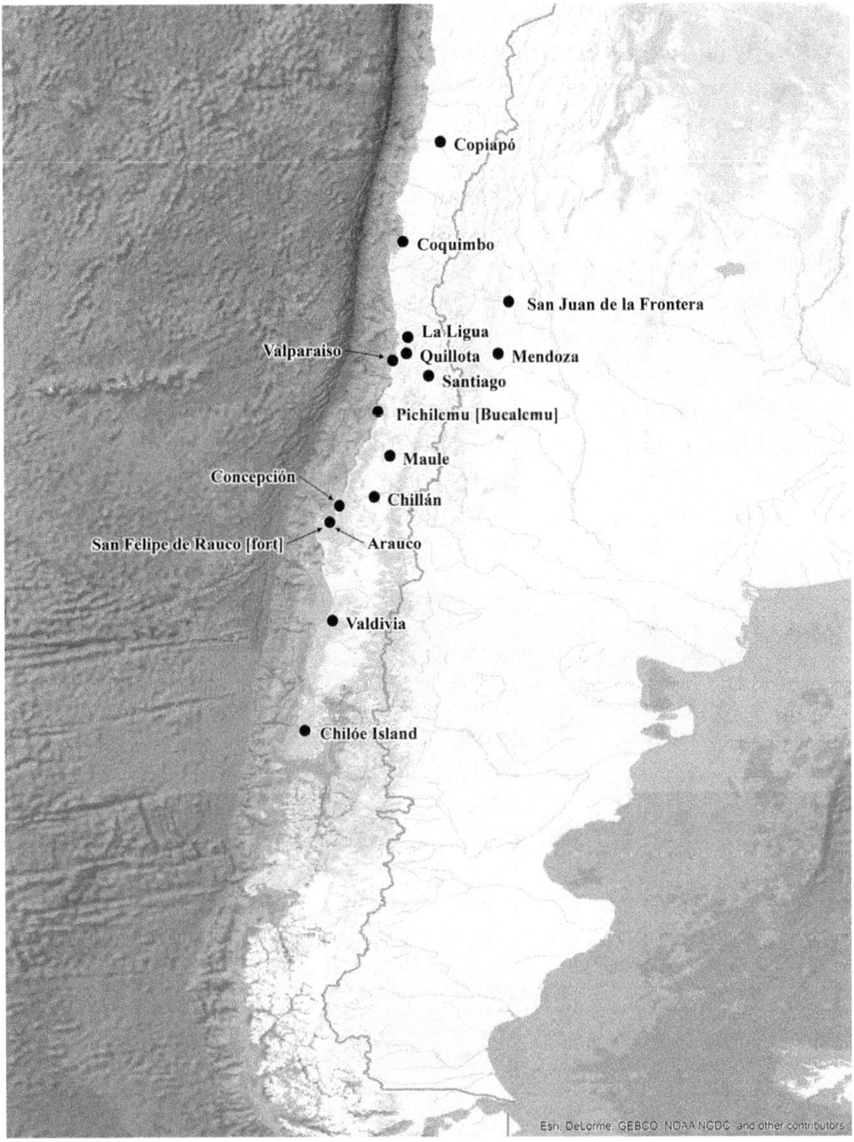

FIGURE 7.8 *Map showing the locations in South America listed on the universal ring dial made for Chile, ca. 1725–1775.*
CREATED WITH THE HELP OF FEI CARNES, CENTER FOR GEOGRAPHIC ANALYSIS, HARVARD UNIVERSITY.

Americans and a few African slaves. In between were those of mixed race. The suitability of this universal ring dial for time finding at so many locations in Chile would have recommended it to an individual of some wealth who had a need to travel—i.e., a royal administrator, a military or naval officer, possibly a merchant, or a missionary.

The most likely user would have been a Jesuit missionary. All of the places listed on the sundial, even the smallest like Bucalemu, have an association with a Jesuit school, *hacienda*, or community. Moreover, the Jesuit theologian, Karl von Haimhausen, the rector of the college in Santiago and confessor to Spanish bishops and the viceroy, was instrumental in establishing in 1748 a significant arts-and-crafts school at Calera de Tango near Santiago. Its workshops were staffed by some fifty German brothers, including a bell-founder, watchmaker, goldsmith, furniture maker, organ builder, painter, sculptor, architect, and printer.[37] Although the universal ring dial was likely made in Spain, it could have been made at this major craft center.

TABLE 7.2 *Gazetteer on a brass universal ring dial designed for use in the Spanish colony of Chile, circa 1750*

Locations on the sundial	Modern place names in Chile, except as noted
Copiapo 25	Copiapó 27° 21' S
Coquimpo 30	Coquimbo 30° S
Ligua 31	La Ligua 32° 27' S
Quillota [32]	Quillota 32° 52' S
S^{n} Juan 32	San Juan de la Frontera [San Juan, Argentina] 31° 32' S
Valparyso [33]	Valparaiso 33° 3' S
Stiago 33.	Santiago 33° 27' S
Maule [34]	Maule 35° 22' S

37 Anthony Huonder, "Karl von Haimhausen," in Charles G. Herbermann et al. (eds.), *The Catholic Encyclopedia*, New York, 1913–1950, 18 vols.; v. 7, pp. 112–113; and Biblioteca Nacional de Chile, *Arte en Chile durante la Colonia. Memoria Chilena*, "Carlos Haimhausen (1692–1767)", www.memoriachilena.cl/602/w3-article-92343.html (accessed 07 March 2016). His name is also spelled Haimbhausen. The author wants to thank her student, Rosario Willumsen of Santiago, for this lead.

Locations on the sundial	Modern place names in Chile, except as noted
Buc^{mu} [34]	Bucalemu [Pichilemu] 34° 23' S
Mend^{za} 34	Mendoza [Argentina] 32° 53' S
Conceptio^{n} [36]	Concepción 36° 49' S
Chillan 36	Chillán 36° 36' S
S^{a} Fee [37]	San Felipe de Rauco [a fort at Old Arauco], 37° 15' S
Arauco 37	Arauco 37° 15' S
Valdiuia 40	Valdivia 39° 48' S
Chiloe 41	Isla Chilóe 42° 41' S

A second extraordinary sundial for use in New Spain was made by Juan Andres in La Paz in 1699. Now part of the collection of the Adler Planetarium, this silver, universal equatorial sundial has eighteen South American places on its gazetteer (*Table 7.3*).[38] These were locations in the northern part of the Viceroyalty of Peru. The sundial does not distinguish between latitudes north or south of the equator, but lists the places in the order of the absolute value of their latitudes (with a few exceptions). The locations include Potosí, known for its great silver lodes and having a population of 160,000 in 1650, making it one of the largest cities in its day in the Western Hemisphere. Lima, the capital of the viceroyalty, was the longtime hub for the export of silver and the import of manufactured goods. Cusco and Quito produced coarse cotton and woolen textiles, while Arequipa and Nazca produced wine.[39] The only ornamentation on this sundial is a simple *globus cruciger* (cross-on-orb), symbolizing the dominion of Christianity over the globe. The decoration suggests that this sundial was made for missionaries, but the locations were also mining, textile, and agricultural centers with European- and colonial-born Spanish administrators and landowners, who oversaw a labor force of enslaved Indians and some Africans.

38 Universal equatorial sundial, Juan Andres, La Paz, 1699, silver, AP inv. A-263.

39 Federal Research Division, Library of Congress, *Peru: A Country Study*, Area Handbook Series, Washington, DC, 1992, pp. 17–24.

TABLE 7.3 *Gazetteer on a silver universal equatorial dial by Juan Andres, La Paz, 1699*

	Locations on sundial	Modern place names
Left column	Quito 0	Quito, Ecuador
	S. Fee 2	[unknown]
	Popaìan 3	Popayán, Columbia
	Paita 5	Paita, Peru
	Zaña 6	Zaña, Peru
	Trugìllo 7	Trujillo, Peru
	Panama 9	Panama City, Panama
	Puertobelo 10	Portobelo, Panama
	Cartagena 10	Cartagena, Columbia
Right column	Lima 12	Lima, Peru
	Nasca 15	Nazca, Peru
	Arequipa 16	Arequipa, Peru
	Guamanga 13	Huamanga [Ayacucho], Peru
	Cusco 14	Cusco, Peru
	La Paz 17	Nuestra Señora de La Paz, Bolivia
	Horuro 18	Oruro, Bolivia
	Plata 19	La Plata [Sucre], Bolivia
	Potosì 20	Potosí, Bolivia

Sundials for Soldiers in North America

The provenances of the Caribbean and South American sundials discussed in the last two sections are lacking for the eighteenth century, leaving us no choice but to make educated guesses about the original users based on the materials, styles, and mathematical projections of the physical objects. By contrast, the pocket sundials considered in this section have better supporting information. These were discovered in archaeological sites at battlefields and forts or have known associations with famous soldiers.

The management of extensive tracts of land in New France and English North America (after 1707, British North America) required the parent nations to build numerous forts to defend their settlements and trading posts. The first fortifications offered protection against Native Americans on whose land the settlers had trespassed. They also prevented European rivals from having access to trading posts, harbors and navigable waterways, and commodities such as fish, timber, and furs. Over the course of the seventeenth and early eighteenth centuries, the British colonies grew in population until British settlers exceeded the French by twenty to thirty times. Less concerned with settlement than trade, the French explored the interior westward from the St. Lawrence River to the Great Lakes and Plains, and southward down the Mississippi to the Gulf Coast. They made alliances with many native groups and their rate of intermarriage was much higher than that of the British.

The animosities between the French and the British culminated in the eighteenth century in a global war fought over colonial supremacy. Known as the Seven Years' War in Europe, the battles started two years earlier in North America, where the war came to be known as the French and Indian War (1754–1763), in recognition that the French were allied with indigenous groups such as the Abenaki, Micmac, Potawatomi, Ojibwa, Ottawa, Shawnee, and Delaware.[40] The Iroquois, who had been traditional enemies of these tribes before European contact, associated themselves with the British.[41] When the war was settled by the Treaty of Paris in 1763, France relinquished control of Canada and its lands east of the Mississippi to Britain in exchange for keeping its sugar-producing colonies in the West Indies. Spain traded Florida to Britain in exchange for Cuba, and received Louisiana from France.

Sundials carried by French officers during the French and Indian War have been recovered near the sites of military campaigns and garrisons during that war. These instruments list the locations of forts and Native American peoples on their gazetteers. Moreover, in keeping with contemporary fashion trends in Paris, the sundials for latitudes north of 40° are found to be the Butterfield-type, complete with the delightful bird-gnomon. One extant sundial intended for use from the Canadian subarctic to Lima (Peru), is an inclining sundial, another common form of the period. A significant group of such sundials made by Pierre le Maire II will be discussed below.

English military officers, on the other hand, preferred more utilitarian compass dials. Like the sundial belonging to Roger Williams, they consisted of a small, shallow, brass box in which a bare magnetic needle rotated on a pivot in the center of a paper wind rose glued to the bottom of the box. The needle and card were protected by glazing held down with a brass frame, which was

---

40 Alan Gevinson, "Which Native American Tribes Allied Themselves with the French?", http://teachinghistory.org/history-content/ask-a-historian/22245, posted under "Ask a Historian", National History Education Clearinghouse, www.teachinghistory.org (accessed 7 March 2016).

41 According to Irene Castle McLaughlin, Curator of North American Ethnography at the Peabody Museum of Archaeology and Ethnology, Harvard University, the French and Indian War can be seen as a conflict of proxies, with Europeans fighting each other through their Indian allies. Personal communication, 11 April 2015.

FIGURE 7.9 *English compass sundial recovered from the site of a British encampment on Rogers Island, adjacent to Fort Edward on the Hudson River, during the French and Indian War.*
FORT TICONDEROGA MUSEUM.

inscribed with an hour scale and supported a gnomon. The gnomon folded flat when the sundial's brass lid was attached. Most compass dials were pocket-sized and barely 4.5 to 5.5 cm in diameter. Each was designed for a particular latitude, but their precision was not high. If a user relocated a couple of degrees north or south, the small size of the sundial kept reading errors in the same limits.

In 1965, an amateur historical archaeologist recovered such a compass sundial (*Fig. 7.9*) in excavations of the former site of an officer's hut on Rogers Island opposite Fort Edward along the Hudson River in New York. This sundial is about 4.5 cm in diameter and has a gnomon for 41° even though the military spot is at a 43° latitude. The compass box of a larger, 7.6 cm-diameter, portable sundial was excavated across the river but nearby at the mainland site of Fort Edward.[42] And just thirty miles away from Fort Edward near Rupert, Vermont, a third English compass sundial was discovered in a buckskin pouch.[43] The district was well-traveled by the British military. Situated at a strategic spot

42 David R. Starbuck, *Excavating the Sutlers' House: Artifacts of the British Armies in Fort Edward and Lake George*, Lebanon, NH, 2010, pp. 38–39 and 41. The partial compass dial is part of the McDougall Collection, owned by the Fort Edward Historical Association. Matthew Keagle, Curator, Fort Ticonderoga, email to author, 29 January 2015.

43 The compass dial in its leather pouch was buried 15 feet below the surface, and discovered during the digging of a building foundation. Ogilvie H. Davis, "Directional Time Tellers of the French and Indian Wars", *Muzzle Blasts* (October 1969), pp. 4–6.

known to Native Americans as the "Great Carrying Place", Fort Edward marked the point on the Hudson River where it became unnavigable going north and travelers had to portage their canoes to Lake George, from where they could access Lake Champlain and continue by water to Canada.[44] Built in 1755, Fort Edward became home to the largest encampment of soldiers—some fifteen to sixteen thousand—during the French and Indian War, and was a staging ground for British campaigns into the Lake George and Champlain valleys in the late 1750s. The Rogers Island compass dial is now on display in the Fort Ticonderoga Museum, and the incomplete dial is held by the Fort Edward Historical Association. The present whereabouts of the sundial discovered near Rupert is unknown.

Archaeologists have also found two gnomons from English compass dials in digs at the Cape Breton site of the Fortress of Louisbourg. Founded by the French in 1713, Louisbourg was a fortified fishing and trading settlement that became home to one of North America's largest fortresses by the mid-1740s. In 1745, British and colonial New England forces captured the fortress during King George's War, but it was returned to France by a peace treaty in 1748. During the French and Indian War, the British regained the Fortress of Louisbourg after a siege in 1758, and its engineers methodically razed it. The British had a garrison there until 1768.[45] The fortress has been reconstructed in modern times, and archaeological excavations done. During the digs, two gnomons from English compass sundials were recovered—one from the site of a storehouse of military and domestic goods and building hardware occupied alternately by French and British forces between 1724 and 1768; the other in a layer of mixed domestic artifacts in a yard similarly occupied between 1727 and 1758.[46] Although Louisbourg is at 45° 54', the first gnomon is from a compass

44 René Chartrand, *The Forts of Colonial North America: British, Dutch, and Swedish Colonies*, Oxford, 2011, pp. 33–35. See the maps: Thomas Jefferys, *The Provinces of New York, and New Jersey; with part of Pensilvania, and the Province of Quebec. Drawn by Major Holland, Engraved by Thomas Jefferys, Geographer to His Majesty. Corrected and Improved, from the Original Materials, By Governr. Pownall, Member of Parliament 1776*, London, 1776: David Rumsey Historical Map Collection, 0346.020; and Claude Joseph Sauthier, *A Map of the Province of New-York, Reduced from the large Drawing of that Province, Compiled from Actual Surveys by Orders of His Excellency William Tryon Esqr. Captain General & Governor of the fame, By Claude Joseph Sauthier; to which is added New-Jersey from the Topographical Observations of C. J. Sauthier & B. Ratzer. Engraved by William Faden, (Successor to the late Mr. Thos Jefferys) 1776*, [n.p.], 1776: David Rumsey Historical Map Collection, 5298.000.

45 René Chartrand, *The Forts of New France in Northeast America, 1600–1763*, Oxford, 2008, pp. 15–16 and 58–59.

46 Gnomon of English compass sundial for latitude 52°, 16L90B5-36, recovered from Archaeological Event 16D19: use of the second storehouse located at the northeast corner

dial made for 52°. This being the typical London product, its owner was probably a British soldier newly transferred to Louisbourg from southern England near the end of the French and Indian War. The second gnomon has an angle of 45°, which indicates that it was made for British forces in North America. It was suitable for use at the garrison at Annapolis Royal, Nova Scotia (44° 50'), which the British seized from the French in 1710, and equally good for use at the nearby Fortress of Louisbourg. It may have been carried from Annapolis Royal to Louisbourg during the British occupation in the 1740s or 1758.

Information about soldiers and their pocket sundials also comes from recollections and estate records. General Edward Braddock, the British commander-in-chief at the start of the French and Indian War, supposedly gave a French silver pocket dial as a sign of friendship to his aide-de-camp, George Washington. At the time a colonel in the Virginia militia, Washington had volunteered his service to Braddock for the ill-fated 1755 expedition to remove French forces from the Ohio territory. When British troops were ambushed and decimated at the Battle of Monongahela, Braddock ordered retreat before being struck by a bullet. With every other officer injured or killed, Washington stepped up to carry out the orders, rallying the forces and employing Braddock's officer's sash as a litter to remove him from the battlefield. As the general lay dying, he gave the sash to Washington.[47] The sash survives, but the story of the pocket sundial is unconfirmed. Another tradition comes from the Revolutionary War period and claims that the Marquis de Lafayette presented a silver pocket sundial to Washington, then commander-in-chief of the Continental Army.[48] As a major general, Lafayette fought in many battles, and Washington treated him as close friend and confident. Whether there is truth in either sundial gift, it is certain that George Washington, like many soldiers, carried pocket sundials

of Lot D (>1724–1768); and gnomon of English compass sundial for latitude 45°, 2L80G12–3, recovered from Archaeological Event 2E23: accumulation of occupation material in the yard of Lot E (1727–1758). These sites existed during the first French period-New England occupation, second French period, and final British occupation of the fortress. Fortress of Louisbourg N.H.S.C., Archaeology Collections. Maura McKeough, Cultural Resource Manager, Cape Breton Field Unit, Parks Canada, email to author, 8 April 2015; and Heidi Moses, Archaeology Collections Manager, Fortress of Louisbourg, email to author, 10 April 2015. The latitude measurements are the author's.

47 Matthew A. Byron, "Battle of the Monongahela," in Joseph F. Stoltz III (ed.), *George Washington Digital Encyclopedia*, Mount Vernon Estate, 2012–, www.mountvernon.org/digital-encyclopedia/article/battle-of-the-monongahela/ (accessed 20 March 2016). Edward Braddock's sprang silk sash, English, c. 1709, Mount Vernon, W-86. On the sundial gift, see Silvio Bedini, *Thinkers and Tinkers: Early American Men of Science*, New York, 1975, p. 264.

48 Alice Morse Earle, *Sun Dials and Roses of Yesterday*, New York, 1902, p. 143.

throughout his military career. Most were of the English compass-dial form. At his death in 1799, an inventory of Washington's possessions at Mount Vernon listed in his study, "1 Pocket Compass" valued at 50 cents, and with valuables in an iron chest, "1 Compass in Brass Case" and "1 Pocket Compass," valued respectively at 50 cents and $ 5.00.[49] Perhaps one of these was the treasured silver pocket sundial allegedly given to him by Braddock or Lafayette.

General Philip Schuyler of Albany fought with the British forces in the French and Indian War and against them in the Revolutionary War. He had a fine brass universal ring dial inscribed with his name. It can be seen at the New York Historical Society.[50] General Jean Baptiste Donatien de Vimeur, comte de Rochambeau, commander of the French troops sent by Louis XVI to aid the American rebels, also carried a pocket sundial.[51] Like Lafayette's sundial, Rochambeau's was undoubtedly an elegant product of a fine Parisian workshop. Its present whereabouts is unknown, but such a silver sundial carried by a French officer in the Revolutionary War survives at the Morristown, New Jersey site of Washington's headquarters and the winter encampment of the Continental Army in 1779–1780, now a National Historical Park. Made in 1673 by Roch Blondeau, Paris, the pocket dial is oval and has hour scales for 40°, 45°, and 50° latitude.[52]

Other sundials recovered from Revolutionary War sites illustrate the preferences of the British forces. A brass compass dial was discovered by archaeologists in 2001 at the Lake George Battlefield Park. The Roman numeral *XXIX* scratched into its case has been interpreted as proof of it being the property of the British 29th Regiment of Foot, which fought at the Battle of Saratoga in 1777. That instrument is now curated by the New York State Museum in Albany.[53] Another compass dial fragment was recovered from a 'dump heap' at

49 Benson J. Lossing, *The Home of Washington; or Mount Vernon and Its Associations, Historical, Biographical, and Pictorial*, Hartford, Conn., 1870, pp. 407 and 409; Bedini, op. cit. (n. 47), p. 264.

50 English universal ring dial, 1750–1780, brass, with a diameter of 11.4 cm, is engraved "Gen. Philip Schuyler" on the equatorial ring. It is in the New York Historical Society, inv. 1941.303, where it is listed wrongly as an "astrolabe"; www.nyhistory.org/exhibit/astrolabe (accessed 8 March 2016). Cf. Bedini, op. cit. (n. 47), p. 264.

51 Earle, op. cit. (n. 48), p. 143.

52 Jude M. Pfister, Chief of Cultural Resources, Morristown National Historical Park, email exchanges with the author, 26 January–3 February, 2015.

53 Starbuck, op. cit. (n. 42), pp. 15–16 and 45–46. The notion that the Roman numerals indicate a connection with a particular regiment deserves further research. Many brass compass dials have Roman numerals scratched on their parts, and these have been presumed to be match marks used in the manufacturing process. See "Compass Dials" in Schechner, *Sundials and Time-Finding Instruments*, op. cit. (n. 11).

Fort Ticonderoga during its reconstruction in the twentieth century. Located at the strategic portage between Lakes George and Champlain, the star-shaped fort was begun in 1755 by the French. Captured by the British in 1759, the stronghold's name was changed from Fort Carillon to Ticonderoga.[54] In 1775, it was the scene of 'America's first victory' in the Revolution, when Ethan Allen, Benedict Arnold, and the Green Mountain Boys captured the fort from the British.[55] The compass dial fragment could have been used by either British or American soldiers (*Fig. 7.10*), whereas a wooden diptych sundial purported to have been carried by Ethan Allen and currently on display at the fort is too late.[56]

Despite wide use of the various forms of pocket sundials by military officers, the historian Silvio Bedini thought none were manufactured in the colonies for this purpose. It is true that surviving examples, inventories, and contemporary records routinely list London and Paris makers.[57] So what do we make of the advertisements of Anthony Lamb in New York City? In 1757, he announced that he made and sold "BATTALIONS of box-wood, according to the new scheme of his royal highness the *Duke of Cumberland*, for distributing the officers proportionably to the several parts of a regiment [...]. Likewise large pocket compasses suitable for the gentlemen of the army".[58] A year later, he advertised "LARGE POCKET COMPASSES, with DIALS, that will serve from 35 to 50 Degrees of Latitude, with the Latitudes of most principal Places in North-America, laid down, suitable for Gentlemen of the Army. Those that have occasion, are desired to apply in Time, lest they be disappointed".[59]

---

54 Chartrand, *The Forts of New France in Northeast America...*, op. cit. (n. 45), pp. 33–37, 57 and 59; Sauthier, op. cit. (n. 44); and Rigobert Bonne, *L'Isle de Terre Neuve, l'Acadie, ou la Nouvelle Ecosse, l'Isle St. Jean et la Partie Orientale du Canada. Par M. Bonne, Ingr. Hydrogr[aphe] de la Marine*, Geneva, 1780; bound into the atlas accompanying Guillaume-Thomas Raynal, *Histoire Philosophique et Politique des Établissemens et du Commerce des Européens dans les deux Indes*, 4 vols., Geneva, 1780: David Rumsey Historical Map Collection, 3083.060.

55 Matthew Keagle, Curator, Fort Ticonderoga, email to author, 29 January 2015. Fort Ticonderoga, *Timeline*, www.fortticonderoga.org/history-and-collections/timeline (accessed 08 March 2016).

56 Thanks to a photo supplied by the Fort Ticonderoga curator, Matthew Keagle, 29 January 2015, I am able to identify the sundial as made by Kleininger in the early 19th century—i.e., after the war and Ethan Allen's death in 1789.

57 Bedini, op. cit. (n. 47), p. 264.

58 Anthony Lamb, advertisement in *The New-York Mercury*, 28 March 1757, issue 242, p. [3], and successive weeks until 25 April 1757.

59 A. Lamb, advertisement in *The New-York Mercury*, 10 April 1758, issue 294, p. [3]; and every week until 19 June 1758.

FIGURE 7.10 *The diversity of soldiers serving in the Continental Army during the Yorktown campaign, 1781, appears in a watercolor sketch made by Jean Baptiste Antoine de Verger, a French sub-lieutenant serving under the Comte de Rochambeau.* From left to right, *de Verger depicted a black soldier of the First Rhode Island Regiment, a New England militiaman, a frontier rifleman, and a French officer. The American patriots favored compass sundials; the French, Butterfield-type or inclining dials.*
ANNE S. K. BROWN MILITARY COLLECTION, BROWN UNIVERSITY LIBRARY.

And in 1760 he announced that his establishment on Hunter's Quay was the place "where Gentlemen of the Army may be supply'd with large Pocket Compasses, with or without Dials, Mahogany, Japann'd and Shagreen Case Telescopes; with all Kinds of Mathematical Instruments in the neatest and best Manner".[60] These advertisements show us that Lamb was savvy enough to take advantage of the business opportunities associated with the French and Indian War. His customers were officers, and they wanted compass sundials that gave both the time and cardinal directions. The sundials Lamb sold that were adjustable for latitude and had gazetteers were probably inclining or Augsburg-type sundials, which contained large central compasses. His customers were likely the same people who would have bought the *New-York Pocket Almanack for the Year, 1757*:

60 A. Lamb, "Suitable for Gentlemen of the Army," *The New-York Mercury*, 7 April 1760, issue 399, p. [1]; S. A. Bedini, *At the Sign of the Compass and Quadrant: The Life and Times of Anthony Lamb*, "Transactions of the American Philosophical Society 74", pt. 1, Philadelphia, 1984.

> Containing, (besides what is usual in other Almanacks) … a List of his *Britannick* Majesty's Land Forces, with the Names of the Field Officers now in *America*, and to what Regiment they belong; the daily Pay of his Majesty's Foot and Marines; a Table shewing how to find the Hour of the Night, by the Shade of the Moon, on a Sun Dial, very convenient for the Gentlemen of the Army.[61]

So why do we not find any extant sundials signed by Lamb? It could be that his sundials were wood and did not endure, that they were unsigned brass instruments and have been mistaken for English products, or that he imported this stock from England. The last option is the most likely when we consider the wide range of goods that he sold, and the regulatory laws that forbade Americans from manufacturing finished goods in British colonies.

The only maker for which there is concrete evidence of his workshop supplying the troops in North America was not American at all. He was Pierre le Maire II of Paris.

Pierre le Maire II, Outfitter of French Officers

Known as Le Maire fils, Pierre le Maire II worked between 1730 and 1760 in Paris at the sign *au Nouveau Quartier Anglais* on the Quai de l'Horloge, Île de la Cité. He was the son and nephew of scientific instrument makers. His father was Pierre le Maire I (b. 1672, fl. 1698–1750), also a maker of sundials but a specialist in mounting lodestones, a point he advertised by working under the sign *à la Pierre d'Aimant*. His uncle was Jacques le Maire (fl. 1714–1762), another sundial maker and member of the Société des Arts. Working under the sign *au Génie*, Jacques had many addresses, but by 1730, father, son, and uncle were all established on the Quai de l'Horloge. Pierre le Maire II worked with his uncle, Jacques, for a time. He was the first maker in France of Hadley's quadrant, which gave meaning to his shop sign, *au Nouveau Quartier Anglais*. He also made sundials, surveying instruments, mathematical instruments, and drawing instruments. Like his father, le Maire was in the Founders' Company, and he served as a juror of masters in 1751–1753. An *ingénieur du Roy*, le Maire made instruments for members of the Académie des Sciences, and was called upon to verify standard measures, like the ell for the city of Paris made by Guibout.[62]

61 *New-York Mercury*, 29 November 1756, issue 225, p. [3].

62 Augarde, op. cit. (n. 23), pp. 54, 62, 67, 71–72; Marcelin, op. cit. (n. 33), s.v. "Lemaire, Jacques", "Lemaire, Pierre Ier", and "Lemaire, Pierre II"; Maurice Daumas, *Scientific Instruments of*

Le Maire's reputation and connections may have led the French government to commission him to supply sundials "to the Officers of their Armies in North America from 1756 to 1759: to assist in guiding them thro[ugh] the Forrests and Wastes of that Country", as a manuscript found in a le Maire sundial case aptly notes (*Pl. 7.10*).[63] Six of these sundials have been located by the author (*Pls. 10–11; Figs. 7.11–13*). Details of their construction and inscriptions are included in *Table 4*. Two of the sundials are in the collection of the Adler Planetarium in Chicago.[64] A third is held by the Musée Stewart in Montreal.[65] The fourth is part of the Peter Winkworth Collection of Canadiana at the Canadian War Museum in Ottawa, which is overseen by the Canadian Museum of History.[66] The fifth is located in Green Bay, Wisconsin at the Neville Public Museum of Brown County.[67] The sixth, and last of the known examples, was in the private collection of Dr. C. H. Lewis of Toronto in the 1960s.[68] Its current location is unknown.

---

the Seventeenth and Eighteenth Centuries and Their Makers (trans. by Mary Holbrook), London, 1972, pp. 182 and 259.

63 Butterfield-type sundial, Pierre le Maire II, Paris, 1730–1750, brass: Musée Stewart, Montréal, inv. 1988.51. The first date has been read both as 1751 and 1756.

64 Butterfield type sundial and inclining sundial, Pierre le Maire II, Paris, 1730–1760, brass: AP, inv. T-58 and W-57. S. J. Schechner, "The Material Culture of Astronomy in Daily Life", *Journal for the History of Astronomy* 32 (2001), pp. 189–222.

65 Butterfield-type sundial, Pierre le Maire II, Paris, 1730–1750, brass : Musée Stewart, Montréal, inv. 1988.51. A. E. Bouchard, "La collection des cadrans solaires du Musée Stewart de Montréal", *Le Gnomoniste* 8, no. 3 (September 2001), pp. 9–13; A. E. Bouchard, "Les cadrans solaires en Nouvelle-France", *Cap-aux-Diamants: La revue d'histoire du Québec* 60 (2000), pp. 18–22, www.erudit.org/culture/cd1035538/cd1044716/7665ac.pdf (accessed 08 March 2016).

66 Butterfield-type sundial, Pierre le Maire II, Paris, 1745–1755, brass: Peter Winkworth Collection of Canadiana, Canadian War Museum, Ottawa, inv. CWM 20080079-011.

67 Butterfield-type sundial, Pierre le Maire II, Paris, 1730–1760, brass: Neville Public Museum of Brown County, Green Bay, Wisconsin, inv. 9184. This sundial was found in 1902 on the eastern shore of Green Bay by a collector, Frank Duchateau, and is sometimes referred to as the Duchateau sundial. It was first published in the *Collections of the State Historical Society of Wisconsin* 16 (1902), opposite p. 64. Brendon Baillod, "New France Sundials from Wisconsin", *Le Gnomoniste* 18, no. 3 (September 2011), pp. 6–7; B. Baillod, "The Mystery of the French Sundials", unpublished MS in the accession records of the Neville Public Museum.

68 Butterfield-type sundial, Pierre le Maire II, Paris, 1730–1755, silver, owned by Daniel-Hyacinthe-Marie Liénard de Beaujeu (1711–1755), whose name is inscribed on the sundial. In the private collection of Dr. C. H. Lewis of Toronto, who purchased it in a London shop in 1960. C. H. Lewis, "Liénard de Beaujeu's compass-sundial", *Antiques* 87, no. 3 (March 1965), "Collectors' Notes", pp. 336–337; Bedini, *Thinkers and Tinkers*, op. cit. (n. 47), figs. 22–23.

FIGURE 7.11 *A rugged wooden case protected this brass sundial as it traveled across the Great Lakes and Plains in New France. Butterfield-type sundial by Pierre le Maire II, Paris, 1730–1760.*
PHOTOGRAPHY BY STEVE PITKIN, PITKIN STUDIO. IMAGE COURTESY OF ADLER PLANETARIUM, CHICAGO, ILLINOIS, T-58.

FIGURE 7.12 *A silver pocket sundial fit for a commander of French forces in Canada, this le Maire Fils sundial, 1730–1750, belonged to Daniel Liénard de Beaujeu. Private collection of Dr. C. H. Lewis, Toronto, 1960s.*
REPRODUCED FROM ANTIQUES 85, MARCH 1965.

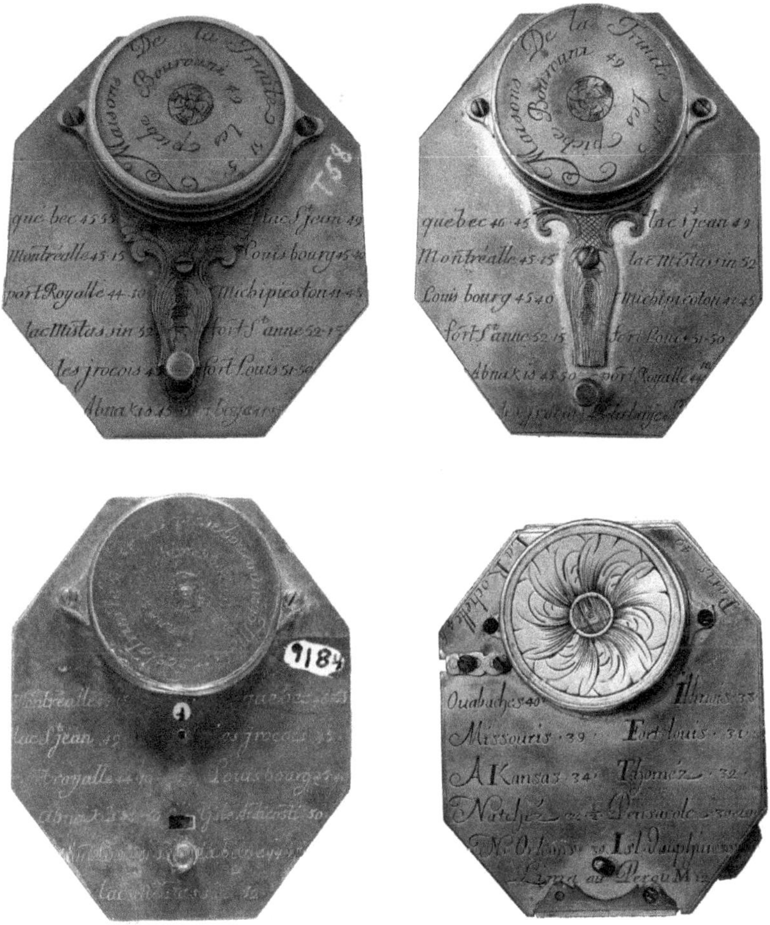

FIGURE 7.13 *Undersides of four le Maire sundials for use in New France. Adler Planetarium T-58, Stewart Museum 1988.5.1, Neville Public Museum 9184, and Adler Planetarium W-57.*
IMAGES COURTESY OF ADLER PLANETARIUM, CHICAGO, ILLINOIS; PHOTOGRAPHY BY STEVE PITKIN, PITKIN STUDIO; STEWART MUSEUM, MONTREAL, CANADA; AND THE NEVILLE PUBLIC MUSEUM OF BROWN COUNTY, WISCONSIN.

TABLE 7.4 *Comparison of le Maire sundials employed in New France*

	Adler T-58	Stewart 1988.5.1	CWM 20080079-011	Neville 9184	Lewis PC	Adler W-57 inclining
Compass box outer ring	Maisons De la Trinité 51. 5'	Maisons De la Trinité 51. 5'	Maisons De la Trinité 51. 5'	Maisons de la Trinité 51. 5' les piche bourouni 49.	Michipicoton 47. 45 Miamis 41. 20 Camanistigoüia 46	
Compass box inner ring	les piche Bourouni 49	Les piche Bourouni 49	Les piche Bourouni 49	fort S^{t} anne 50. 15' Michipicoton 41. 45'	Chagoüamigon 46 Oüisconsin 42. 30	
Left column	qué bec 45. 55 Montréalle 45. 15 port Royalle 44. 10 lac Mistassin 52 les jrocois 45 Abnakis 45. 50'	québec 46. 45' Montréalle 45. 15' Louis bourg 45. 40' fort S^{t} anne 52. 15' Abnakis 45. 50' les jrocois 45.	québec 46. 45' Montréalle 45. 15' Louis bourg 45. 40' fort S^{t} anne 52. 15' Abnakis 46. 50' les jrocois 45.	Montréalle 45. 15' lac S^{t} jean 49. port royalle 44. 10' abnakis 45. 50' fort Louis 51. 50' lac Mistassin 52.	Québec 46. 55 Trois-Rivieres 46. 20 S^{t}. Frédéric 44. 10 Niagara 42. 55 Nipigon 50 S^{t}. Joseph 42. 30 Missilimakinac 45. 9	Ouabaches . 40. Missouris . 39. AKansas . 34. Natchéz 32. 1/2. N. Orleans . 30. Lima au Perou . M. 12.

	Adler T-58	Stewart 1988.5.1	CWM 20080079-011	Neville 9184	Lewis PC	Adler W-57 inclining
Right column	lac S^{t}jean 49 Louis bourg 45. 40' Michipicoton 41. 45' fort S^{t}anne 52. 15' fort Louis 51. 50' la baye 44. 15'	lac S^{t}jean 49. lac Mistassin 52. Michipicoton 41. 45' fort Louis 51. 50' port Royalle 44. 10' la baye 44. 15'	lac S^{t}jean 49. lac Mistassin 52. Michipicoton 41. 45' fort Louis 51. 50' port Royalle 44. 10' la baye 44. 15'	québec 46. 45' les jrocois 45. Louis bourg 45. 40' l'jsle Auticosti 50. la baye 44. 15'	Montréal 45. 15 Chambli 45. 15 Frontenac 44. 10 Le Détroit 42. 10 Chicagou 42. La Baye 44. 15	Illinois . 38. Fort . Louis . 31 . . Thoméz . 32. Pensacole . 30. 20' Isl. dauphine 30
Left side compass box					Nouvelle	La Rochelle 46
Right side of compass box					France	Paris 49.
Signature	P^{re} . le Maire AParis	P^{re} . le Maire AParis	P^{re} . le Maire AParis	P^{re} . le Maire AParis	Le Maire fils AParis	P. Le Maire / AParis
Hour scales	43, 46, 49, 52	43, 46, 49, 52	43, 46, 49, 52	43, 46, 49, 52	43, 46, 49, 52	Inclining
Material	Brass	Brass	Brass	Brass	Silver	Brass
Other	Rugged wooden field case	Original Paris case with manuscript label			*Inscribed around compass box:* LIENARD DE BEAUJEV	

Five are octagonal, Butterfield-type dials, complete with the adorable bird on the gnomon for setting the latitude of the sundial between 40° N and 60° N. Four of these are made of brass; one of silver. Each sundial has four hour scales, and every scale is divided from 4 am to 8 pm, with quarter-hour subdivisions. The innermost hour scale is for 43° N latitude, and successive scales are for 46°, 49°, and 52°. On the brass sundials, the hours alternate between Roman and Arabic numerals: 43° is marked V-XII-VII; 46°, 4–12–8; 49°, IIII-XII-VIII, and 52°, 4-12-8. On the silver sundial, all hour scales are marked in Roman numerals. Each sundial has an inset magnetic compass with a wind rose labeled in French and a fleur-de-lys at north. On the brass sundials, the wind rose has sixteen simple points identified by initials, whereas the silver sundial has thirty-two points, sixteen of which are split triangles in the nautical style. Its cardinal and intercardinal points are engraved with their full French names; the remaining points are labeled by French initials. When the compass needle is present on the brass dials, it is a blued-steel rod intersected by a circle near the northern end. The silver sundial, however, has a blued needle shaped as an arrow with a feathered tail (*Fig. 7.12*).[69] The compass boxes are offset about 14°–15° W, which matches the magnetic declination in Paris around 1730–1740 as well as a swathe through Nova Scotia and Canada. Additional ornamentation includes acanthus leaves on the hour plate at the toe of the gnomon and alongside the bird index on the gnomon, and a hint of an architectural scroll supporting the hour scales. The ornamentation is more artfully engraved and refined, of course, on the silver sundial. The brass sundials are signed *Pre. le Maire AParis*. The silver example is signed, *Le Maire fils AParis*. The silver sundial is also distinguished by being inscribed *Nouvelle France* (New France) and *LIENARD DE BEAUJEV* (the name of its famous owner, Daniel-Hyacinthe-Marie Liénard de Beaujeu).

The sixth sundial (*Pl. 11*) in this special group is a brass inclining dial signed underneath the hour plate *P. Le Maire / AParis*. Its magnetic compass is similar to those of the brass Butterfield-type dials already described, except that it is offset for a magnetic declination of about 16° 45'–17° W, which was appropriate for Paris circa 1750. The single hour scale is likewise marked IIII-XII-VIII and divided into quarter hours. The folding latitude arm is divided 50° N-0°-40° S, every 1° and inscribed *Partie Septentrionalle* (northern part) and *Partie Meridionalle* (southern part).

The top sides of these six sundials are constructed and ornamented in the style typical of many other mid-eighteenth century French sundials in general,

69 Compass Needle Register, types 10 and 66, respectively, prepared by the author for the forthcoming sundial and time finding instrument catalogue of the Adler Planetarium.

and those of Pierre le Maire in particular. It is only when they are turned over that we see how unusual they are. Instead of the usual gazetteer of French towns and major European cities, we discover a list of sites in North America (*Fig 7.13, Table 7.3*).

The sundial currently at the Neville Public Museum was unearthed in 1902 from a newly plowed field on the eastern shore of Green Bay, Wisconsin, an arm of Lake Michigan. The field was near Red Banks, the site of a large palisaded village of the Winnebago in 1634 when the French explorer, Jean Nicolet, arrived by canoe in search of resources and a shortcut to China. Nicolet set up a trading post there and called the place *La Baye des Puants* (the bay of the stinking). The *puants* referenced were the Sioux-speaking people that he found living in the village, whose Algonquian-speaking neighbors called *winpyeko* (Anglicized to Winnebago), meaning 'people of the dirty water'. This referred to the muddy water of the lower course of the Fox River (linking Lake Winnebago to Lake Michigan at Green Bay), which became clogged every summer with rotting fish.[70] La Baye des Puants appears on early maps of New France, such as that drawn by Louis Joliet in 1673–1674, but was frequently shortened to *Puans* or *La Baye*.[71] In 1669 a Jesuit father arrived in La Baye and established a mission. A fortified trading post followed in 1684, and was rebuilt in 1717, when a garrison was stationed there. Fort La Baye remained in French hands until the British captured it in 1761 during in the French and Indian War (*Fig. 7.14*).[72] The sundial was probably lost by one of the last French officers to occupy the post between 1756 and 1758, but it may also have belonged to a missionary.[73]

70 Nancy Oestreich Lurie, "Winnebago," in William C. Sturtevant et al. (eds.), *Handbook of North American Indians*, Washington, DC, 1978–2008, 20 vols.; v. 15, pp. 690–707; Charles M. Rousseaux, "Winnebago," in Sharon Malinowski and Anna Sheets (eds.), *The Gale Encyclopedia of Native American Tribes*, Detroit, 1998, 4 vols.; v. 3, pp. 330–335.

71 Louis Joliet, *Nouvelle Decouverte de Plusieurs Nations Dans la Nouvelle France, En L'année 1673 et 1674*, MS map: John Carter Brown Library, Providence, RI, Map Collection, Cabinet B674 / 1 Ms.

72 Jacques Nicolas Bellin, the royal hydrographer, shows the fort and mission clearly on the site in his 1764 map, *La Nouvelle France où Canada*, Paris, 1764: David Rumsey Historical Map Collection, 6903.012. On Fort Labaye, see R. Chartrand, *The Forts of New France: The Great Lakes, the Plains, and the Gulf Coast, 1600–1763*, Oxford, 2010, pp. 28–29.

73 Baillod, "The Mystery…", op. cit. (n. 67). A reference to a Jesuit father having a sundial is found in a letter from La Ronde, commandant at Fort Chequemagon, to Beauharnois, 22 July 1738, in which the writer confirms a report that Sioux had killed twenty-two Frenchmen, saying that it "was quite true, for we have had Sun-dials and several other things belonging to the Jesuit Father". See *Collections of the State Historical Society of*

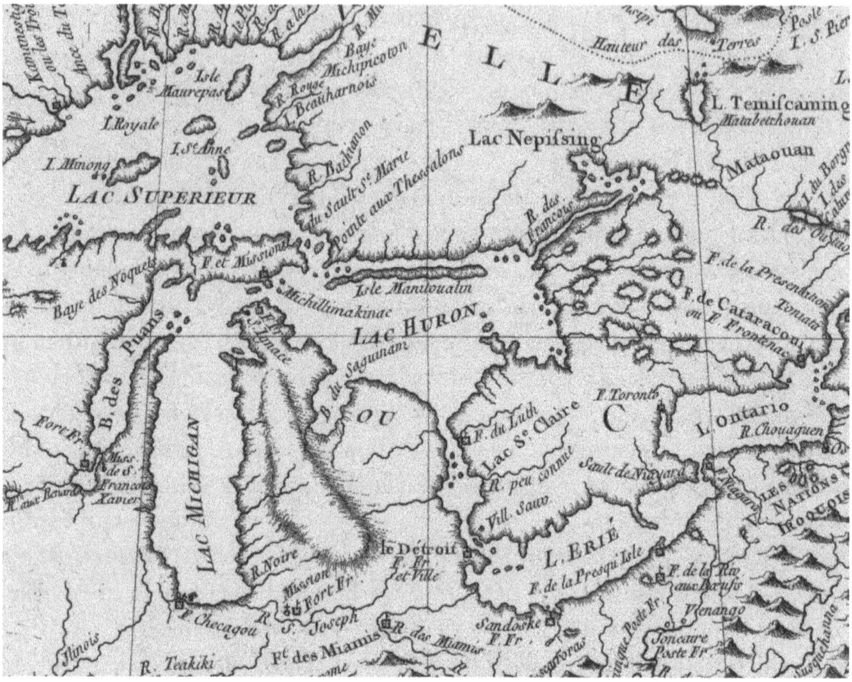

FIGURE 7.14 *Jacques Nicolas Bellin, the royal hydrographer, clearly shows* La Baye des Puans *with its mission and fortified trading post in the lower left of this detail of his 1764 map. The map also shows forts and Indians around the Great Lakes that were listed on the le Maire sundials, including Kaministiquia, Michipicoten, Michilimackinac, Frontenac, the Iroquois, Niagara, Detroit, Miamis, and Chicago.*

BELLIN, LA NOUVELLE FRANCE OU CANADA, *PARIS, 1764: DAVID RUMSEY HISTORICAL MAP COLLECTION, 6903.012.*

Fort La Baye (*La Baye 44. 15*) appears on every one of the le Maire Butterfield-type sundials. Other sites listed on the gazetteers were significant outposts in New France. *Québec 45. 55* referred to Quebec City, established by Samuel de Champlain in 1608. On the St. Lawrence River, it became the main trading port and the administrative capital of New France. Montreal (*Montréalle 45. 15'*), the principal interior city in the province of Quebec was established in 1642 as a colonial mission, but quickly became a major fur trading post. Its early years were marked by a series of conflicts with the Iroquois Confederacy, known as

---

Wisconsin 17 (1906), pp. 277–278, along with a photo of a bronze garden sundial recovered near Portage, Wisconsin in 1903.

the Iroquois Wars or the Beaver Wars.[74] On the northwestern coast of what is today Nova Scotia, Port Royal (*port Royalle 44. 10*) was the capital of the French colony of Acadia from 1605 until 1710, when the British seized it and renamed it Annapolis Royal.[75] Notice that the sundials did not concede the defeat. French and Acadian forces, allied with the Micmac, attempted to recapture the town four times during King George's War (1744–1748). In 1755 the British began deporting the Acadian people from Annapolis Royal because they wanted to cut off their support to the nearby Fortress of Louisbourg (listed on the sundials as *Louis bourg 45. 40'*).

Other forts on these eighteenth century French sundials also show the tug-of-war between France and Britain in North America. *Fort St Anne 52. 15'* was first known as Fort Albany in the 1670s when the Hudson's Bay Company had established it as a fur trading post and a defense on James Bay at the mouth of the Albany River. The French captured the post in 1686 and renamed it Fort Sainte Anne. The English recaptured it in 1693 and reverted the name. Today it is the home of the Fort Albany First Nation and is accessible only by air or ice roads. *Fort Louis 51. 50'* was another fur trading post established in 1671 by the Hudson's Bay Company as a stockade under the name of Moose Fort. In 1686, the French captured Moose Fort and renamed it Fort Louis. Ten years later in 1696, the English retaliated and burned it to the ground. In 1730 the Hudson's Bay Company set up a new fort a mile upstream. Today the site is called Moose

74 Paul-André Linteau, "Montréal" and "Iroquois Wars", *The Canadian Encyclopedia*, Toronto, 2009, www.thecanadianencyclopedia.ca/en/article/montreal/, and Anonym, www.thecanadianencyclopedia.ca/en/article/iroquois-wars/ (accessed 9 March 2016).

75 Nicolas Landry and Anselme Chiasson, "History of Acadia", ibid., www.thecanadianencyclopedia.ca/en/article/history-of-acadia/ (accessed 9 March 2016); Chartrand, *The Forts of New France in Northeast America* . . ., op. cit. (n. 45), pp. 12–16. Compare names on French and British maps: it is Port Royal on Guillaume de Lisle, *Carte du Canada ou de la Nouvelle France et des Decouvertes qui y ont été faites. Dressée sur plusieurs Observations et sur un grand nombre de Relations imprimées ou manuscrites. Par Guillaume De l'Isle, Geographe de l'Academie Royale des Sciences*, Paris, 1703 [between 1708 and 1712?]: David Rumsey Historical Map Collection, 4764.097; and it is Annapolis Royal on Herman Moll, *New Found Land, St. Laurence Bay, the fishing banks, Acadia, and part of New Scotland*, London, 1736: David Rumsey Historical Map Collection, 5580.047. Both names are given together on Didier and Gilles Robert de Vaugondy, *Partie de l'Amérique Septent*[*rionale*] *qui comprend la Nouvelle France ou le Canada, par le Sr. Robert de Vaugondy Géog*[*raphe*] *Ordinaire du Roy*, Paris, 1755: David Rumsey Historical Map Collection, 3353.099; and Bonne, op. cit. (n. 54).

Factory, Ontario. Maps of the eighteenth century give both English and French names for each of these sites.[76]

Since 1714, Michipicoten (*Michipicoton 41. 45'*) was the site of a fur trading post on the northeastern shore of Lake Superior at the mouth of the Michipicoten River. The post was at the junction of the two main trade routes—one westward from Montreal and the other north to James Bay. The name is Ojibwa for 'big bluffs'. It also refers to the Michipicoten people whose traditional summer grounds were at the mouth of the river.[77] The French first visited the area between 1617 and 1622, and Michipicoten appeared on Samuel de Champlain's map of 1632 and other later maps.[78]

Maisons de la Trinité 51. 5' may have been settlements near the Monts de la Trinité, a mountain range just north of the Manicouagan River. In 1664, Henri Nouvel was the first Jesuit missionary to travel upstream on the river to Manicouagan Lake, which was formed from an impact crater. Along the way, he celebrated mass on the festival of the Most Holy Trinity within sight of a great mountain. This being the "first sacrifice ever offered in this country, where no European had been before", he named the peak Mont de la Trinité.[79] Although *Mont de la Trinité* is found on contemporary maps, as well as a plural form (*Mts de la Trinité*) applied to a range of mountains, no site identified as *Maisons de la Trinité* has been found on maps or in the *Jesuit Relations*.[80] The label on the sundial may be a corruption of *Monts de la Trinité*. There is also another possibility: in France, Maisons de la Trinité (Trinity Houses) were relief organizations for sailors and their widows in the mid-eighteenth century, and this listing on the sundials' gazetteers may have referred to such establishments

76 Chartrand, *The Forts of New France in Northeast America…*, op. cit. (n. 45), pp. 54–55; Bellin, op. cit. (n. 72).

77 Michipicoten First Nation, "Who We Are", www.michipicoten.com/who-we-are/ (accessed 09 March 2015).

78 Samuel de Champlain, *Carte de la nouvelle France, augmentée depuis la derniere, servant a la navigation faicte en son vray Meridien, par le Sr. de Champlain Capitaine pour le Roy en la Marine; lequel depuis l'an 1603 jusques en l'année 1629*, in S. de Champlain, *Les Voyages de la Nowelle France occidentale, dicte Canada*, Paris, 1632: Bibliothèque Nationale de France, Département Cartes et Plans, GE C-6108 (RES). See also de Lisle, *Carte du Canada…*, op. cit. (n. 75); Bellin, op. cit. (n. 72); and H. Moll, *A New Map of ye North Parts of America claimed by France under ye Names of Louisiana, Mississipi, Canada & New France. with the Adjoyning Territories of England & Spain*, London, 1736: David Rumsey Historical Map Collection, 5580.046.

79 Reuben Gold Thwaites (ed.), *The Jesuit Relations and Allied Documents: Travels and Explorations of the Jesuit Missionaries in New France, 1610–1791*, Cleveland, 1896–1901, 73 vols.; v. 49, pp. 46–47 (1664).

80 For the plural form, see for example, Bonne, op. cit. (n. 54).

along the St. Lawrence River. We must remember that the sundials reflect the decisions of administrators in Paris about what was important in New France, whether or not this was the case on the ground in Canada. Bigger forts and fur trading posts such as Michilimackinac were overlooked on the sundials, and smaller ones such as La Baye and Michipicoten were included.[81]

By the latitude given on the sundials, *Lac Mistassin 52* would appear to be now known as Mistassin Lake, Ontario. However, the largest freshwater lake in Quebec and a degree further south, Lake Mistassini, was frequently identified in seventeenth- and eighteenth-century French accounts and maps as Lac Mistassin and other variations of the Cree word for 'large rock'. It was an important waypoint for traders traveling up the Saguenay and Rupert Rivers to James Bay.[82]

Along the same trading route was *Lac St Jean 49*, which also had a French settlement and fur trading post. The local people, the Kakouchak Innu, called it *Piekouagami*, or 'flat lake'. Some French maps denote a people by the name *Piekouagamiens* near the lake, and it at first seems that the sundial's *Les piche Bourouni 49* may be a much altered spelling given the latitude assigned on the sundials.[83] However, other maps show a people called the *Pitchibourouni* living further north on the eastern coast of Hudson Bay (*Fig. 7.15*).[84] The name refers to a native band living in what is now East Main Cree territory. In the *Jesuit Relations*, they were variously called the Kilistinons Nisibourounik, the Pishhapocanoes "near akin to the Eskeimoes", the Pitchibourenik, and Pitchibourouni.[85] The sundial maker's source has conflated the two.

Anticosti Island (*l'île Auticosti 50*), is strategically situated where the Saint Laurence River enters the gulf. Originally a hunting ground for native peoples, Louis XIV presented the island to Louis Joliet in 1680. He built a fort and lived with his family there. Only one of the sundials includes this destination.[86]

The remaining geographic locations on the brass Butterfield-type sundials are references to territories inhabited by two native tribal groups: the Abenaki (*Abnakis 45. 50'*) and the Iroqouis (*les irocois 45*). The Abenaki were Algonquian

81 Chartrand, *Forts of… the Great Lakes…*, op. *cit.* (n. 72), p. 5; Baillod, "The Mystery…", op. cit. (n. 67).

82 De Lisle, *Carte du Canada…*, op. cit. (n. 75); Bellin, op. cit. (n. 72); Bonne, op. cit. (n. 54).

83 The Native American name for the lake (Piekouagami) is given alongside the French name (Lac St. Jean) on the maps by de Vaugondy, op. cit. (n. 75), and Bonne, op. cit. (n. 54). Bellin, op. cit. (n. 72) shows the people (Piekouagamiens) as well.

84 De Lisle, *Carte du Canada…*, op. cit. (n. 75).

85 Thwaites (ed.), op. cit. (n. 79), v. 44, p. 248 (1658); v. 45, p. 228 (1660); v. 56, p. 202 (1672); Richard J. Preston, "East Main Cree", in Sturtevant et al. (eds.), op. cit. (n. 70), v. 6, pp. 196–205, esp. p. 205.

86 Neville Public Museum inv. no. 9184.

FIGURE 7.15 *The Pitchibourouni people—listed on the le Maire sundials as* Le piche Bourouni*—lived on lands east of Hudson Bay. Detail of Guillaume de Lisle,* Carte du Canada ou de la Nouvelle France et des Decouvertes qui y ont été faites, *Paris, 1703.*
DAVID RUMSEY HISTORICAL MAP COLLECTION, 4764.097.

speakers (*Fig. 7.16*). Their name comes from their own language and means 'dawn land people' or 'easterners'. Their territory extended across New England, southern Quebec, and the Canadian Maritimes. The Abenaki were strong allies with the French.[87] Their traditional enemies before European contact were the Iroquois, a confederacy of five tribes whose languages were part of the same

87 Dean R. Snow, "Eastern Abenaki", in Sturtevant et al. (eds.), op. cit. (n. 70), v. 15, pp. 137–147; Geoffrey Parkinson, "Abenaki", in Malinowski and Sheets (eds.), op. cit. (n. 70), v. 1, pp. 7–14.

FIGURE 7.16 *Abenaki couple, eighteenth century watercolor.*
COURTESY OF THE CITY OF MONTREAL RECORDS MANAGEMENT & ARCHIVES, MONTREAL, CANADA.

language family and whose territories were located in upstate New York and Canada. The tribes included the Mohawk, Oneida, Onondaga, Cayuga, and Seneca. A sixth tribe, the Tuscarora, joined the Iroquois confederacy in 1722. The Iroquois called themselves *Haudenosaunee* (People of the Long House) in reference to the construction of their homes. It was the French who labeled them *Iroquois*, deriving it from a pejorative Algonquian name meaning 'real adders'. French alliances with the Abenaki strengthened their animosity with the Iroquois. The British invested in the Iroquois, and used them as proxies to fight native groups allied with France during the French and Indian War.[88] The sundials, thus, informed users of the locations of friends and foes, but made these Indian nations seem smaller by confining them to very specific latitudes.

Exceptional among these Butterfield-type sundials is the elegant silver one owned by Daniel-Hyacinthe-Marie Liénard de Beaujeu (1711–1755) (*Fig. 7.17*).[89] Born in Montreal, Liénard de Beaujeu was an officer in the French colonial troops like his father, Louis Liénard de Beaujeu. One of the posts where his father served was Fort Michilimackinac, a regional command center and a key trading post on the Straits of Mackinac, the waterway that connects Lake Huron and Lake Michigan.[90] As a lieutenant, Liénard de Beaujeu was among the troops that tried to recapture Louisbourg and Port Royal from the British in 1746. In 1749, he became commanding officer at Fort Niagara, located at the mouth of the Niagara River on Lake Ontario. It had a sizeable stone edifice intended to withstand British artillery.[91] In 1755, Liénard de Beaujeu was sent to command the newly built Fort Duquesne, situated at the meeting point of the Allegheny and Monongahela rivers (in present day Pittsburgh, Pennsylvania). He and his men set out from Lachine near Montreal on the 500 mile journey. En route he stopped at Fort Frontenac, Fort Niagara, Fort de la Presqu'île, Fort de la Rivière au Bœuf, and so forth, building a supply line until he reached Fort Duquesne.[92] Warned of the approach of General Braddock's troops, Liénard de Beaujeu organized a band of French and Indian fighters (Ottawa, Delaware, Huron, and Abenaki) to ambush the British troops. The maneuver was successful, but he died at the start of the battle.

His sundial (*Fig. 7.15*) tells the story of his life, listing the significant French forts of *Montréal 45. 15*, *Missilimakinac 45. 9*, *Niagara 42. 55*, and *Frontenac 44. 10*. It also included Quebec (*Québec 46. 55*) and Trois-Rivières (*Trois-Rivières 46. 20*), the first and second permanent settlements in New France, established in 1608 and 1634 respectively. Of further interest to a soldier like Liénard de Beaujeu were Fort Detroit (*Le Détroit 42. 10*), established in 1701; Fort Saint-Frédéric

88 William N. Fenton, "Northern Iroquoian Culture Patterns", in Sturtevant et al. (eds.), op. cit. (n. 70), v. 15, pp. 296–297; Loretta Hall, "Iroquois", in Malinowski and Sheets (eds.), op. cit. (n. 70), v. 1, pp. 75–84; Correspondence between the author and Irene Castle McLaughlin, Curator of North American Ethnography at the Peabody Museum of Archaeology and Ethnology, Harvard University, April 2015.

89 Malcolm MacLeod, "Liénard de Beaujeu, Daniel-Hyacinthe-Marie", in *Dictionary of Canadian Biography*, Toronto, 2003–2016, v. 3 (1741–1770), www.biographi.ca/en/bio/lienard_de_beaujeu_daniel_hyacinthe_marie_3E.html (accessed 9 March 2016).

90 Chartrand, *Forts of… the Great Lakes…*, op. cit. (n. 72), pp. 4–5 and 27–31.

91 Ibid., pp. 4–6 and 18–22.

92 On these significant military posts, see ibid., pp. 4–8, 15–21, 25.

FIGURE 7.17 *Chevalier Liénard de Beaujeu (1711–1755), owner of the silver Le Maire sundial. From Benjamin Sulte,* Histoire des Canadiens-français, 1608–1880, *Montreal, 1882–1884, v. 5, 6th illustration preceding the title page.*
COURTESY OF THE BIBLIOTHÈQUE ET ARCHIVES NATIONALES DU QUÉBEC, 52327/1956461.

(*St. Frédéric 44. 10*), established in 1734 on Lake Champlain; Fort Kaministiquia (*Camanistigoüia 46*), rebuilt in 1717 on Lake Superior at Thunder Bay, Ontario, and its northern dependency, Fort Nipigon (*Nipigon 50*), a fur trading post; Fort Saint Joseph (*St. Joseph 42. 30*), a fur trading post set up in 1691 on the St. Joseph River at Niles, Michigan; Fort Chambly (*Chambli 45. 15*) established in 1711 on the Richelieu River in Quebec, Canada; and Fort Chequamegon (*Chagoüamigon 46*), a stockaded trading post since 1661 on a small bay off Lake Superior, and home to fur traders, copper miners, and garrisons since 1692.[93] His sundial also drew his attention to the southern end of the Wisconsin River from the point where it joined the Mississippi and continued eastward for about a hundred miles (*Oüisconsin 42. 30*). At this eastern point, travelers could portage their canoes to the Upper Fox River and head downstream to Lake Winnebago and the Lower Fox River, which emptied into Green Bay. This was part of an important waterway that linked the Mississippi to the Great Lakes.[94] Another strategic landmark was Chicago (*Chicagou 42.*), which gave southern Lake Michigan access to the Gulf of Mexico via the Chicago River and a portage to the Des Plaines River, which joined the Illinois River going downstream to the Mississippi.[95] Chicago was also a trading center for the Potawatomi, Miami, and Illinois. The Miami, an Algonquian-language people living in what is now Indiana, southwest Michigan, and western Ohio, may have earned their own spot on the sundial (*Miamis 41. 20*), but this inscription, being so specific in latitude, more likely refers to Fort Saint-Philippe des Miamis. Built in 1715 where the St. Mary's River and St. Joseph River joined to form the Maumee River and near a large Miami town, the fort defended trade routes, but was attacked several times by hostile native groups. A British garrison took charge in 1760, and today it is Fort Wayne, Indiana.[96]

---

93 Ibid., pp. 4–8, 15–34. These forts are shown on maps such as de Lisle, *Carte du Canada*…, op. cit. (n. 75); de Vaugondy, op. cit. (n. 75), and Bellin, op. cit. (n. 72).

94 The waterway and portage is clearly noted on the map of Nicolas de Fer, *Le Cours du Missisipi, ou de St. Louis: Fameuse Riviere de l'Amerique Septentrionale aux Environs de laquelle se trouve le Païs appellee Louisiane*, Paris, 1718: David Rumsey Historical Map Collection, 10022.100; and G. de Lisle, *Carte de la Louisiane et du Cours du Missisipi. Dressée sur un grand nombre de Memoires entr'autres sur ceux de Mr. le Maire. Par Guillme. de l'Isle de l'Academie Rle. des Sciences*, Amsterdam, 1742: David Rumsey Historical Map Collection, 4638.095.

95 De Lisle, *Carte de la Louisiane*…, op. cit. (n. 94); de Vaugondy, op. cit. (n. 75).

96 Charles Callender, "Miami", in Sturtevant et al. (eds.), op. cit. (n. 70), v. 15, p. 681. The Miami people are featured on many maps of the period, such as de Fer, op. cit. (n. 94); Bellin, op. cit. (n. 72); de Lisle, *Carte de la Louisiane*…, op. cit. (n. 94); de Vaugondy, op. cit. (n. 75). Chartrand, *Forts of… the Great Lakes*…, op. cit. (n. 72), pp. 25–26.

While these Butterfield-type dials took their owners across French Canada, the inclining sundial by le Maire took its owner down the Mississippi and its tributaries to the Gulf of Mexico (*Fig. 7.18*). Most of the listings on its gazetteer referred to native peoples, rather than the French forts near them. Starting in the north, the sundial called attention to the Wabash Confederacy (*Ouabaches . 40.*), Native Americans of different tribes who lived along the Wabash River in present-day Illinois, Indiana, and Ohio. Their designation came from a Miami name for the river, *Wah-bah-shi-ki* (meaning pure white, bright), which referred to the water flowing over the river's limestone bed. French Jesuits shortened the name.[97] The Missouri (*Missouris . 39.*) were a Southern Sioux nation encountered by Jacques Marquette and Louis Joliet in 1673 and noted on their map as *Oumessourit*. They lived along the Missouri River and were known as 'people of the dugout canoes', a name given them by the Illinois nation in their Algonquian language.[98] The Illinois appeared on the sundial (*Illinois . 38.*) as well. Comprised of many related tribes (*Pl. 12*), this large nation historically occupied the central Mississippi River valley (running through the states of Illinois, Iowa, Missouri, and Arkansas).[99] The Akansa (*AKansas . 34.*) referred to the Quapaw tribe who lived where the Arkansas River met the Mississippi. The group was part of the Dhegiha Sioux, which split into several tribes that left the Ohio valley. In moving down the Mississippi River, the Quapaw became known as the *Ogaxpa*, or 'downstream people'. Marquette, the French explorer, first made contact with them in 1673. His Illinois guides, however, called the Quapaw the 'people of the south wind', or the Akansa. They were on friendly terms with the French.[100] Natchez (*Natchéz 32. 1/2.*) was the location of the primary ceremonial mounds and Grand Village

97 Thwaites (ed.), op. cit. (n. 79), v. 59, p. 145 (1673–1675); U.S. Geological Survey Geographic Names Information System, http://geonames.usgs.gov/apex/f?p=gnispq:3:0::NO::P3_FID:449850 (accessed 9 March 2016).

98 K. Michelle Moran, "Missouri," in Malinowski and Sheets (eds.), op. cit. (n. 70), v. 3, pp. 302–306. The Missouri are shown on N. de Fer, *Partie Meridionale de la Riviere de Missisipi, et ses environs, dans l'Amerique Septentrionale. Mis au jour par N. de Fer, Geographe de sa Majesté Catolique*, Paris, 1718: David Rumsey Historical Map Collection, 10022.099.

99 Charles Callender, "Illinois", in Sturtevant et al. (eds.), op. cit. (n. 70), v. 15, pp. 673–680; Greg Mazurkiewicz, "Illinois," in Malinowski and Sheets (eds.), op. cit. (n. 70), v. 1, pp. 69–74. See also Illinois State Museum, "The Illinois Indians", www.museum.state.il.us/muslink/nat_amer/post/htmls/il.html (accessed 9 March 2016). The Illinois appear prominently on maps such as de Fer, *Partie Meridionale de la Riviere de Missisipi*, op. cit. (n. 98).

100 Thwaites (ed.), op. cit. (n. 79), v. 65, pp. 113–125 (1700); v. 69, p. 217 (1750); de Fer, *Partie Meridionale de la Riviere de Missisipi*, op. cit. (n. 98); Candace LaBalle, "Quapaw",

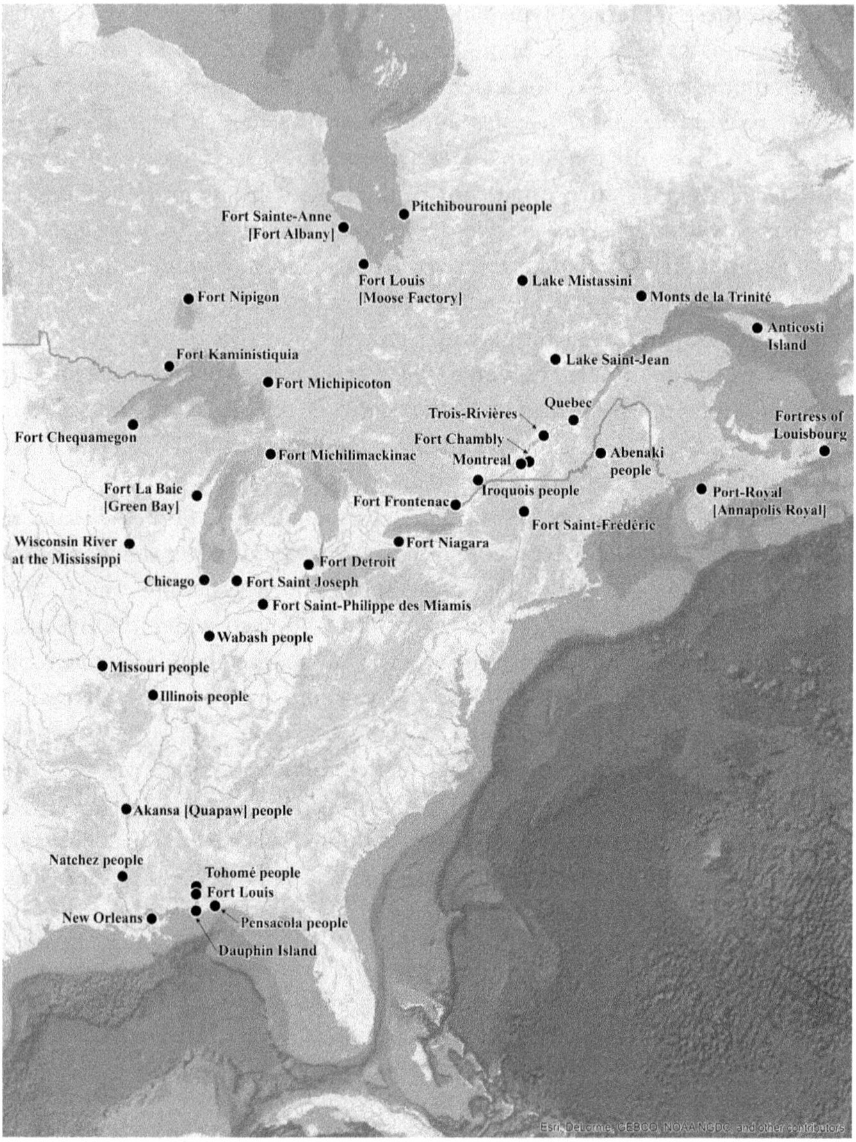

FIGURE 7.18 *Map showing the locations in North America listed on the sundials made by Pierre le Maire II for use in New France, 1730–1760.*
CREATED WITH THE HELP OF FEI CARNES, CENTER FOR GEOGRAPHIC ANALYSIS, HARVARD UNIVERSITY.

of the Natchez tribe along the Mississippi River. The French established a trading post there in 1713, built Fort Rosalie in 1716, and created a colony in 1718. When a despotic commander at the fort demanded in 1729 that the Natchez people leave their sacred Grand Village to make way for his plantation, the people rebelled. The retaliation was harsh. In 1731, the French soldiers forced about four hundred Natchez to surrender, burned some in a public spectacle, and sold the rest into slavery in Saint Domingue.[101] Lastly the sundial identified the city of New Orleans (*N. Orleans . 30.*) at the mouth of the Mississippi. Founded in 1718 by the French Mississippi Company, New Orleans became the capital of Louisiana in 1722 (*Fig. 7.19*).

Further east along the gulf, the Pensacola tribe (*Pensacole . 30. 20'*) lived near modern-day Pensacola, Florida. The Spanish established a garrison there in 1698. In 1719, the French took charge, but surrendered to the Spanish in 1722. The Pensacola were at war with the indigenous people living along the Mobile River to the west. These included the Mobila and Tohomé. The Tohomé were also listed on the sundial (*Thoméz . 32.*), and were allied with the French colonists settled on the Mobile River at La Mobile (today Le Moyne, Alabama).[102] La Mobile was the capital of French Louisiana from 1702 to 1711. It was guarded by Fort Louis (*Fort-Louis . 31.*) on Mobile Bay. In 1699, the French explorer Pierre Le Moyne d'Iberville had visited Mobile Bay and established a colony on 'Massacre Island', which was renamed Dauphin Island (*Isl. Dauphine 30*) in 1707.[103]

---

in Malinowski and Sheets (eds.), op. cit. (n. 70), v. 3, pp. 350–355. The official website for the tribe today is www.quapawtribe.com (accessed 9 March 2016).

101 De Fer, *Partie Meridionale de la Riviere de Missisipi*, op. cit. (n. 98); Patricia Galloway and Jason Baird Jackson, "Natchez and Neighboring Groups", in Sturtevant et al. (eds.), op. cit. (n. 70), v. 14, pp. 598–615; Doris Morris Maxfield, "Natchez", in Malinowski and Sheets (eds.), op. cit. (n. 70), v. 1, pp. 463–467; Chartrand, *Forts of... the Great Lakes...*, op. cit. (n. 72), pp. 48–50.

102 George E. Lankford, "Chacato, Pensacola, Tohomé, Naniaba, and Mobila", Sturtevant et al. (eds.), op. cit. (n. 70), v. 14, pp. 664–668; Frederick Webb Hodge, *Handbook of American Indians North of Mexico*, "Smithsonian Institution, Bureau of Ethnology 30", Washington, DC, 1907; John R. Swanton, *The Indian Tribes of North America*, "Smithsonian Institution, Bureau of American Ethnology 145", Washington, DC, 1952.

103 Chartrand, *Forts of... the Great Lakes...*, op. cit. (n. 72), pp. 53 and 57–58; and maps by N. de Fer, *Les Costes aux Environs de la Riviere de Misisipi. Decouvertes par Mr. de la Salle en 1683. et reconnues par Mr. le Chevallier d'Iberville en 1698. et 1699. Par N. de Fer, Geographe de Monseigneur le Dauphin*, Paris, 1701: Bibliothèque Nationale de France, Département Cartes et Plans, GE D-13313; de Fer, *Partie Meridionale de la Riviere de Missisipi*, op. cit.

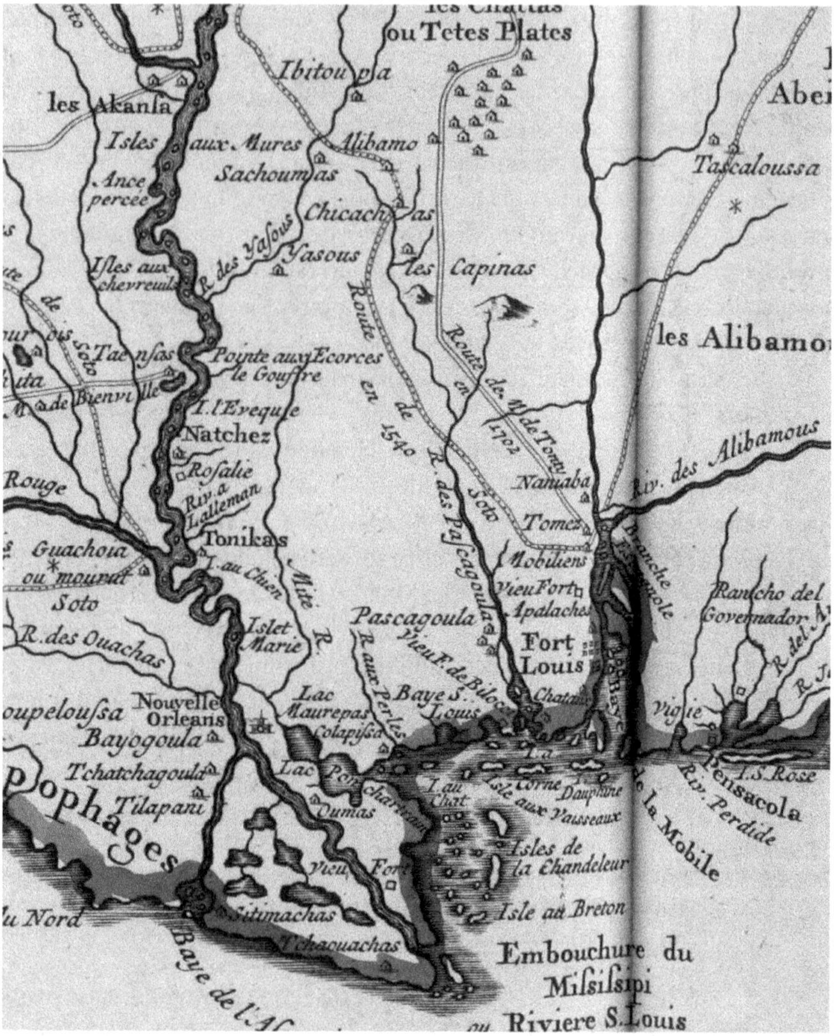

FIGURE 7.19 *The Akansa and Natchez people along the Mississippi River, the Tohomé on the Mobile River, and the Pensacola along the Gulf coast are identified on this map, along with Fort Rosalie, New Orleans, Fort Louis, and Dauphin Island. Guillaume de Lisle,* Carte de la Louisiane et du Cours du Mississipi, *Amsterdam, 1742, detail.*
DAVID RUMSEY HISTORICAL MAP COLLECTION, 4638.095.

(n. 98), which shows the tribal villages very clearly; de Lisle, *Carte de la Louisiane…*, op. cit. (n. 94); MOLL, *A New Map of ye North Parts of America…*, op. cit. (n. 78); and Emanuel Bowen, *A New & Accurate Map of Louisiana, with Part of Florida and Canada*, London, 1747: David Rumsey Historical Map Collection, 3733.058.

As for the other cities on this sundial, the importance of *Paris 49.* needs no discussion. Lima, Peru (*Lima au Perou . M. 12.*) was of interest to the French after the transfer of the Spanish Crown to the House of Bourbon in 1700. And *La Rochelle 46*? This western coastal city of France was active in the triangular trade of slaves from Africa, sugar from plantations in the West Indies, and fur from Canada. In 1718, the French West India Company sent eight hundred colonists from La Rochelle to Louisiana. Among them was Antoine-Simon le Page du Pratz. He was appointed superintendent of the public plantations, but is now better known for his ethnographic account of Louisiana, which provided much information about the lives of native peoples and colonists at the time this sundial was made in Paris.[104]

Conclusions

What can we conclude from all these remote forts, tribal lands, trading posts, sugar plantations, and colonial administrative seats inscribed in brass, silver, and ivory? What do we make of sundial relics plowed up on old battlefields or discovered in collections? These tangible things shed light on what French, British, and Spanish administrators, instrument makers, and users thought most important to have and identify in the vastness of colonial North and South America. In a European city, for instance, a sundial's magnetic compass was a mere accessory for orienting the instrument to find time. In the forests and plains of the Americas, however, the compass became essential for finding one's way, while the hour scale was secondary. Perhaps this was why British officers were willing to carry single-latitude compass sundials whose time-finding accuracy was never very good. Their choice also shows us that "gentlemen of the military" of British and French backgrounds favored different styles of sundial in the field, just as they did back home in Europe. In addition, a hierarchy of class and rank is manifest in the materials employed and the quality of the engraving on the sundials. The extant dials, by virtue of the latitudes for which they were made, also enable us to gauge the variety and geographical range of potential users. These included plantation owners, wealthy merchants, military officers, colonial officials, and missionaries,

104 Antoine-Simon le Page du Pratz. *Histoire De La Louisiane, Contenant La Découverte De Ce Vaste Pays; Sa Description Géographique; Un Voyage Dans Les Terres; L'histoire Naturelle, Les Moeurs Coûtumes & Religion Des Naturels, Avec Leurs Origines; Deux Voyages Dans Le Nord Du Nouveau Mexique, Dont Un Jusqu'à La Mer Du Sud; Ornée De Deux Cartes & De 40 Planches En Taille Douce*, Paris, 1758.

all of whom were predominantly white males, but could also have been prosperous men and women who were freed black slaves or individuals of mixed race.

The sundials are also evidence of the transmission of cartographic and ethnographic knowledge gathered by missionaries, explorers, soldiers, and colonial merchants. At the same time, the dials testify to the wish of the colonial administrators to circumscribe and minimize the boundaries of the aboriginal populations by limiting their ranges to single latitudes on the instruments. The sundials, moreover, illustrate the tug-of-war of opposing imperial powers in their assignment of names to wilderness places and indigenous peoples. French and English names were swapped back and forth for settlements and geographical features, and the labels of native groups were often slurs imposed on them by invaders who were led by native guides belonging to tribal enemies of the people being classified. More disturbingly, the sundials also demonstrate a zeal for marketable commodities and military might by individuals and governments who were willing to enslave, resettle, and massacre Africans and Native Americans in their pursuit.

Lastly, the European sundials destined for colonial use surprise us by their practicality; especially as material evidence that Europeans wanted to have time discipline in a wilderness that did not run on Western notions of time.

CHAPTER 8

Selling Mathematical Instruments in America before the Printed Trade Catalog

*Richard L. Kremer**

Introduction

In 1849, America's first advertising agent, Volney B. Palmer, sponsored an essay contest on the "subject of advertising". The winning entry, penned by Horace Greeley, editor of New York's largest newspaper, presented a "philosophy of advertising" for the new age of enterprise. "Man, they say, is a trading animal—the only one. The wants of each individual are more numerous and varied than his faculties and capacities devoted to supplying them". According to Greeley's economic history of the world, trade arose to meet these wants. To succeed, the trading merchant must not only be able to satisfy demand but also to insure that "everybody within the proper scope of his business is made aware of his ability and confident of his disposition to do so". Failure to advertise, Greeley warned, signals one's "gross incompetency and ruinous prodigality [...] [and] bestows the spoils on his wiser rivals". To neglect advertising, "is like resolving never to travel by steam nor communicate by telegraph. It is to close one's eyes to the light, and insist on living in perpetual darkness".[1]

Palmer's and Greeley's campaign, historians of American advertising have suggested, signaled the emergence, in the 1840s, of a new discourse and practice of advertising in American print culture. Before that, advertising could appear unseemly, as if the advertiser were a quack, pretender or huckster, overly anxious for business or seeking unscrupulous profits. In an era of the

* I wish to thank the editors and referees of this volume for their many helpful suggestions; Philip Bell, Jonathan W. Chipman, Gloria Clifton, Paul Musselwhite, Roger Sherman, Steven Turner, and Deborah J. Warner for advice on particular topics; and Jane Carroll and my student assistant, Audrey McCartney, for helping with research from afar. I also thank staff at the New London County Historical Society, Massachusetts Historical Society, Boston Public Library, Yale's Beinecke Library, and Dartmouth College's Rauner Special Collections Library for kindly assisting my research in their collections.

1 Horace Greeley, "The philosophy of advertising", *V. B. Palmer's business-men's almanac*, New York, 1851, pp. 48–50. Cf. Donald P. Holland, "Volney B. Palmer: The nation's first advertising agency man", *Pennsylvania magazine of history and biography* 98 (1974), pp. 353–381.

 | DOI 10.1163/9789004324930_009

political press, many colonial and early national newspapers defended partisan causes. Paid advertising, viewed as patronage for these political publishers, could seem antidemocratic, especially after 1776. And the mercantile press, generally marked by the term "advertiser" in its titles, often provoked moral censure for "soliciting" advertisements from all manner of disreputable individuals or enterprises.[2]

The goal of this essay is not to explore the moral valence of advertising in early American culture. Rather it seeks to examine how one particular American craft, the makers and retailers of mathematical instruments, sought to sell their products in the eighteenth century. Did they advertise? How did they advertise, and to whom? Did they define the 'proper scope' of their business, to quote Greeley, as the local town, region, colony, or after 1776, the nation? How do these marketing efforts elucidate the development of what by the 1830s would be called 'scientific instrument making' on the American continent? And what do the advertising strategies tell us about the cultural significance of numeracy, precision, and 'science' in a frontier far from European centers of instrument making such as London or Paris? Can we discover any differences between American advertising strategies and those pursued by the far better studied British instrument makers in the eighteenth century?

To explore these questions, we shall first survey the makers of mathematical, optical and philosophical instruments in America during the colonial and early national periods (constraints of length require me to halt the analysis in 1800). I will then review several previous studies of British instrument makers' advertising strategies before analyzing a sample of about 200 newspaper advertisements and considerably fewer printed trade cards and broadsides prepared by American makers between 1707 and 1800. We will conclude that most eighteenth-century American makers did not advertise consistently over their careers; that makers in the larger towns and cities advertised more than those in rural areas; and that, judging by the content of their advertisements, American makers expected to sell primarily to local customers. In conclusion, we shall examine several makers' account books to see whether their sales

2 For an introduction to this massive literature, see Stephen R. Fox, *The mirror makers: A history of American advertising and its creators*, New York, 1984; Richard M. Ohmann, *Selling culture: Magazines, markets, and class at the turn of the century*, New York, 1996; Tim P. Vos and You Li, "Justifying commercialization: Legitimating discourses and the rise of American advertising [1800–1870]", *Journalism & mass communication quarterly* 90 (2013), pp. 559–580. For a useful, comparative study of the development of "newspaper cultures" in eighteenth-century Britain and America, see Uriel Heyd, *Reading newspapers: Press and public in eighteenth-century Britain and America*, Oxford, 2012.

were indeed concentrated within a few miles of their shops. It is Greeley's "proper scope" for American mathematical instrument making that guides our investigation.

Mathematical Instrument 'Makers' in America to 1800

A longitudinal survey of the development of the making and selling of mathematical, optical and philosophical instruments in eighteenth-century North America, more particularly of the makers, importers, and sellers of these products, remains a desideratum for the history of science and technology.[3] To get a rough sense of the spread, through space and time, of individuals in North America who represented themselves as making and selling instruments, I have assembled a database of names, locations and dates by consulting several earlier studies and various relevant databases (see *Appendix*). Although surely incomplete and less thorough than the British directories, my preliminary compilation may nonetheless reveal some trends and hopefully will prompt further research on the early American trade in mathematical instruments. My preliminary database contains 180 names of makers, importers, or retailers who had commenced their trade in North America by 1800.[4] Given the small, family-based workshop structure of the trade, I would guess that a count of firms or businesses would yield roughly the same number.

Note that I place the term 'maker' in quotation marks in the heading to this section. Just because men signed their advertisements as 'makers' does not guarantee that they were constructing apparatus from wood, brass, and glass; likewise, even signed instruments might well have been made in London and imported to America, where a retailer would add his name. As we shall see below, many American 'makers' undoubtedly only repaired and sold

3 If we follow actors' categories, traffic in "scientific" instruments would not begin until the 1830s, as demonstrated in a now classic study by Deborah J. Warner, "What is a scientific instrument, when did it become one, and why?", *British journal for the history of science* 23 (1990), pp. 83–93. For directories of British instrument makers, see E. G. R. Taylor, *The mathematical practitioners of Tudor and Stuart England, 1485–1714*, Cambridge, 1954; E. G. R. Taylor, *The mathematical practitioners of Hanoverian England, 1714–1840*, Cambridge, 1966; Gloria Clifton, *Directory of British scientific instrument makers, 1550–1851*, London, 1995; A. D. Morrison-Low, *Making scientific instruments in the Industrial Revolution*, Aldershot, 2007. For North America, the only related study remains James M. Edmonson, *American surgical instruments: The history of their manufacture and a directory of instrument makers to 1900*, San Francisco, 1997.

4 I have found another 360 names for the years 1800 to 1860.

instruments. Economic historians have stressed the structural impediments to colonial manufacture in British America such as small markets compared to British makers producing for an entire empire, high labor costs where land is cheap, and minimal division of labor.[5] My preliminary database thus records men who represented themselves as 'makers'; considerably more research would be needed to sort out who actually made what. Hence, in what follows readers should think 'maker' wherever they see the term maker.[6]

As Deborah J. Warner has noted, many of the earliest makers in North America emigrated from England, where they had learned the trade. Anthony Lamb (1703–1784), in the earliest known American newspaper advertisement for a mathematical instrument maker, described himself in 1730 as "from London, now living in New-York". John Dabney, Jr. announced in his 1739 Boston advertisement that he had "serv'd his time to Mr. Jonathan Sisson [London], mathematical instrument maker to His Royal Highness, the Prince of Wales". Charles Walpole's 1740 Philadelphia advertisement indicates that he was a "mathematical instrument maker from London".[7] As we shall see below from my survey of the eighteenth-century advertisements, however, most American makers did not publicly announce the particulars of their training or national origin. More prosopographical research would be required to explore this issue in detail.

My sample of 180 makers suggests, first and not surprisingly, that instrument making began in America's largest cities. The earliest known maker, James Halsey (d. 1696), is identified in a Boston land deed of 1674 as a "mathematician" and by the inscription on what is apparently the earliest known instrument made in America, a Davis quadrant in pear wood: "MADE BY JAMES HALSEY FOR THOMAS SMITH 1676".[8] After 1700, several other Halseys would open instrument-making shops in Boston. In 1704, Christopher Witt immigrated from Wiltshire, England to Germantown (just north of Philadelphia), where he became known locally as a physician, botanist,

5 John J. McCusker and Russell R. Menard, *The economy of British America, 1607–1789, with supplementary bibliography*, Chapel Hill, NC, 1991, pp. 309–330. Carole Shammas, *The pre-industrial consumer in England and America*, New York, 1990, p. 68, estimates that in 1774 the average American was spending about 30% of his or her budget on imported goods.

6 I thank Sara Schechner and the referees of this paper for underscoring this point.

7 All quoted from D. J. Warner, "British instrument makers in the U.S.", *Rittenhouse* 6 (1991), pp. 25–31.

8 D. J. Warner, "Davis' quadrants in America", *Rittenhouse* 3 (1988), pp. 23–40; Marion Vernon Brewington, *The Peabody Museum collection of navigating instruments with notes on their makers*, Salem, 1963, pp. 6–7 and 128.

mystical Pietist, clockmaker and maker of mathematical instruments, including telescopes.[9] The earliest known maker in Philadelphia proper was Thomas Godfrey (1704–1749), a friend of Benjamin Franklin and in 1730 inventor of an improved Davis quadrant.[10] The above mentioned Anthony Lamb had been apprenticed to the London mathematical instrument maker Henry Carter in 1717 and would become New York's first maker, after being transported in 1724 to the colonies as a convicted criminal.[11]

TABLE 8.1 *Estimated white population in eighteenth-century British America, in thousands*

	1700	1730	1760	1790
All colonies	275	655	1610	3930
Philadelphia	4	8	19	42
New York	5	9	14	33
Boston	7	13	16	18

Boston, Philadelphia, and New York would remain the three most populous cities in eighteenth-century North America (see *Table 8.1*),[12] and instrument making would remain prominent in those cities. My database finds 28 makers in Boston through 1800, 27 in Philadelphia and 19 in New York, 7 in Baltimore, 5 in Newport, 4 in Salem, and 2 in Charleston. These six cities were the only large urban areas in colonial America (with populations exceeding 8,000 at

9 Julius Friedrich Sachse, *The German Pietists of provincial Pennsylvania*, Philadelphia, 1896, pp. 402–418; Edward W. Hocker, "A doctor of colonial Germantown: Christopher Witt, physician, mystic and seeker after the truth", *Germantown history* 2, no. 8 (1948), pp. 205–226; Silvio A. Bedini, *Early American scientific instruments and their makers*, Washington, D.C., 1964, pp. 62–63.

10 James Logan, "An account of Mr. T. Godfrey's improvement of Davis's quadrant", *Philosophical transactions* 38 (1734), pp. 441–450; Bedini, op. cit. (n. 9), p. 58.

11 For a richly detailed account, see S. A. Bedini, "At the sign of the compass and quadrant: The life and times of Anthony Lamb", *Transactions of the American Philosophical Society* 74, no. 1 (1984). Lamb had fallen in with one of London's most notorious burglars.

12 U.S. Bureau of the Census, *A century of population growth from the first census of the United States to the twelfth, 1790–1900*, Washington, D.C., 1909, pp. 1–15. My study does not include the West Indies; I have found only one maker in that region (see *Appendix*).

some time); roughly half of the makers in my database worked in these cities. The other half worked in another 54 smaller towns and villages. Demographers have estimated that between 1730 and 1820, no more than 5% of the white population in colonial and early federal America lived in the six large urban areas; indeed over the course of the eighteenth century, America deurbanized, with a smaller proportion of the whites living in towns in 1775 than had lived there in 1700.[13] Clearly, the instrument makers did not simply distribute themselves evenly among the white population of the American continent; they remained overly represented in urban and coastal locales.[14]

Second, the non-urban makers tended to congregate, as can be seen in *Pl. 13* and *Table 8.2*,[15] in the Connecticut River Valley, a region of innovative craft and proto-industrial production (furniture, clocks, guns) by the 1750s, and in the middle colonies of Pennsylvania and Maryland. Strikingly, almost no makers are found in the southern colonies. Was instrument making, already in the eighteenth century, starting to concentrate in areas where industrialization would emerge in the early nineteenth century? Does the greater per capita wealth (for free white persons) in the southern colonies suggest that they purchased their mathematical instruments directly from London makers without seeking the services of local makers?[16] Did the South, distinctly more rural than the other colonies, simply lack the urban concentrations where makers tended to congregate?[17]

13 Shammas, op. cit. (n. 5), p. 53.

14 Morrison-Low also found that instrument making in provincial England did not simply track population growth but rather took root in given cities for a variety of reasons. See Morrison-Low, op. cit. (n. 3), pp. 38–40.

15 Makers from *Appendix*, population data from the first American census, conducted in 1790; see Susan B. Carter et al., *Historical statistics of the United States, Millennial edition online*, Cambridge, 2006, http://hsus.cambridge.org/HSUSWeb/toc/hsusHome.do (accessed 12 March 2016). I have combined the state-based census data to match the original thirteen British American colonies. The 1790 census did not include Native Americans.

16 According to McCusker and Menard, op. cit. (n. 5), p. 61, wealth per free white person, ca. 1770–1775, was £ 33 in New England, £ 51 in the middle colonies, and £ 132 in the southern colonies.

17 Russell R. Menard, "Economic and social development of the South", in Stanley L. Engerman and Robert E. Gallman (eds.), *Cambridge economic history of the United States, vol. 1: The colonial era*, Cambridge, 1996, pp. 249–295: 252.

TABLE 8.2 *Distribution of eighteenth-century American instrument-makers by colony, population in thousands*

	Total makers	1790 white population	1790 black population
New Hampshire, Vermont	13	227	1
Massachusetts, Maine	47	476	6
Rhode Island	8	65	4
Connecticut	15	232	5
New York	21	314	26
New Jersey	4	170	14
Pennsylvania	41	423	10
Delaware	2	46	13
Maryland	13	298	111
Virginia, W. Virginia, Kentucky	5	504	318
N. Carolina & Tennessee	0	321	110
S. Carolina	3	140	109
Georgia	0	61	13

Third, instrument making remained a very specialized and small trade within the economic activity of the colonies and the new nation. The earliest Boston city directory (1789), "a list of the merchants, mechanics, traders and others", includes 1,486 names, mostly men but a few women (widows), and their street addresses and occupations. The frequency distribution of those occupations reveals a city of merchants, shopkeepers, and retailers, with a considerable spread of crafts, especially in the building and shipbuilding trades (see *Table 8.3*).[18] Twelve clock- or watchmakers are listed, but only three mathematical instrument makers.

18 Ann Smith Lainhart, *First Boston city directory (1789), including extensive annotations by John Haven Dexter (1791–1876)*, Boston, 1989, pp. 2–4 and 111–112.

TABLE 8.3 *Leading Boston occupations in the 1789 city directory*

Merchant	121
Shopkeeper	88
House wright	72
Tailor	66
Storekeeper	44
Boarding house operator	40
Cooper	33
Retailer	30
Baker	29
Hair dresser	25
Blacksmith	25
Shoemaker	25
Physician, surgeon	23
Mariner	21
Printer, stationer, bookseller	21
Brick layer, mason	20
Distiller	20
Innholder	18
Carpenter	17
Shipwright	16
School teacher	16
Attorney at law	15
Apothecary	12
Goldsmith	12
Watch or clock maker	12
Mathematical instrument maker	3

Finally, my database reveals a period of considerable growth in the number of new instrument makers in America from 1740 through 1800. By 1740, 15 makers are known; by 1760, 1780 and 1800, respectively, an additional 25, 45 and 92 names enter my database. These increases are exponential, with a doubling time of about 23 years, a rate of growth in instrument makers that is somewhat higher than that found in England over this same period.[19] More

19 Cf. Clifton, op. cit. (n. 3), p. xv; Morrison-Low, op. cit. (n. 3), p. 45. My preliminary database, which extends to 1860, suggests that the rate of growth in American makers slows somewhat after 1800.

research would be required to fully explain this rapid growth, which must have been driven by consumer demand. We might speculate that American customers increasingly wanted locally made (less expensive?) apparatus rather than London exports, especially as the War of Independence approached. Or we might guess that demand in general increased as the colonies expanded. A. D. Morrison-Low has suggested that the eighteenth-century market for instruments included three groups: men of science (philosophers) who sought apparatus for their investigations of the natural world; dilettanti or gentlemen with resources to buy instruments to amuse or entertain their friends and family; and mathematical practitioners who needed instruments for utilitarian work such as surveying, navigating or measuring in a practical trade.[20] As we shall see below, the advertisements provide some clues about the composition of the American market in the eighteenth century.

The Rhetoric and Strategies of Advertising Instruments in Eighteenth-century America

Since many of the early mathematical instrument makers in colonial America came from England, as did the early printers and publishers, we might expect that advertising practices in colonial America would closely follow patterns previously established in London. We begin, then, with a brief survey of English advertising of instruments before 1800.

The earliest London advertising for mathematical instruments appeared, already in the 1550s, in books on surveying, navigation, or instruments, in what historian David J. Bryden has called "informal" advertisements or brief references buried in the texts to local makers who could supply the item being discussed. Informal advertisements may not have been instigated by a maker or retailer but merely provided by an author as a helpful recommendation for his readers. What Bryden describes as "formal" advertisements, set apart by format and typography from their carrying media (books, almanacs, newspapers), began to appear after 1600. Both types, however, rarely list prices, seldom specify the range of particular products being offered, and, Bryden speculates, probably were aimed at "professional" practitioners who knew what a given category (e.g., carpenters' rules, quadrants, dials) entailed and would cost. Sixteenth- and seventeenth-century London makers probably

20 Ibid., pp. 27–29.

worked mostly on commission, providing bespoke instruments to mathematical practitioners.[21]

Only in the eighteenth century, as optical and philosophical instruments joined the mathematical, did the London market broaden to clients with "leisurely" rather than "professional" interests. Craftsmen turned to the media of trade cards, broadsides and handbills, positioning their products amidst "luxury" goods and services such as spices, silks and fine linens, silver shoe buckles, fashionable furniture, playing cards, etc., depicted in baroque—and even rococo—styled engravings.[22] The palette of available instruments became ever more diverse. By the 1740s, the first published catalogues appeared, listing in several pages literally hundreds of instruments and signaling a broader commercialization of the trade. For example, in 1746, the London maker, George Adams, included a 20-page "Catalogue of mathematical, philosophical and optical instruments, as made and sold by George Adams, at Tycho Brahe's Head in Fleet Street" at the end a lengthy, popular book on microscopy that he authored and published. Clearly aimed at a wide market, Adams numbered each instrument in the catalogue (335 in total) "so that if a Gentlemen is desirous of any one or more of them, and is at any distance from London, he need only send me the numbers adjoining to those he intends to purchase, and he shall be served with fidelity, and at the lowest prices" (no prices are listed).[23] In 1765, a similar 14-page catalog by the London makers Heath and Wing would be appended to John Hammond, *The practical surveyor*, a pedagogical text

21 D.J. Bryden, "Evidence from advertising for mathematical instrument making in London, 1556–1714", *Annals of science* 49 (1992), pp. 301–336. For a specialized set of related early advertisements, see Sarah Tyacke, *London map-sellers, 1660–1720: A collection of advertisements for maps placed in the London Gazette, 1668–1719 with biographical notes on the map-sellers*, Tring, 1978.

22 H.R. Calvert, *Scientific trade cards in the Science Museum collection*, London, 1971; Michael A. Crawforth, "Evidence from trade cards for the scientific instrument industry", *Annals of science* 42 (1985), pp. 453–554; Philippa Hubbard, "Trade cards in 18th-century consumer culture: Movement, circulation, and exchange in commercial and collecting spaces", *Material culture review* 74/75 (2012), pp. 30–46; Maxine Berg and Helen Clifford, "Selling consumption in the eighteenth century: Advertising and the trade card in Britain and France", *Cultural and social history* 4 (2007), pp. 145–170.

23 George Adams, *Micrographia illustrata, or the knowledge of the microscope explain'd*, London, 1746, p. 244. The categories in this catalogue include instruments for geometry, drawing; rulers of all sorts; surveying; navigation; instruments for shewing the motion, attraction, weight, and equilibrio of bodies; experiments on the motion, weight, and equilibrio of fluids; pneumatical experiments; optical instruments; experiments on lights and colours; experiments on electricity; instruments for astronomy, geography.

aimed at mathematical practitioners. Interestingly, the Heath and Wing catalog divides its products into categories identical to those in Adams's list.[24] By 1750, London's mathematical instrument makers were advertising via informal notices in books, formal advertisements in newspapers, trade cards and broadsides, and multi-paged catalogs appended to practical manuals. By 1800, provincial English makers had followed suit. As Morrison-Low wrote: "provincial instrument businesses grasped every new method of promoting their wares as soon as they could: establishing attractive shop windows in fashionable spas; issuing literature which drew attention to patronage by the nobility and gentry; advertising the availability of a wide variety of items, whether described in famous textbooks or made to order; selling by lottery; adding price lists to the backs of textbooks; participating in exhibitions of an educational nature; and undertaking lecture tours using demonstration apparatus".[25] Would instrument makers in the American colonies employ similar strategies?

The earliest American advertisement I have found appeared not in a newspaper but in an annual almanac, another periodical publication that widely circulated in colonial America, starting already in the 1660s. Samuel Clough (1665–1707), who authored almanacs printed in Boston for the years 1700–1708, signed himself a "lover of mathematics" on his title pages. At the end of his imprint for 1707 he announced his offer of "all sorts of money scales & weights, large and small, made & sold by the author, at his shop at the Sign of the Hand & Scales" (see *Fig. 8.1*).[26] The next year's almanac, appearing after Clough had died, carried a notice by his widow that their shop again offered "money seals [sic] and weights" made by "an ingenious young man who was an apprentice to her late husband".[27] I have found nothing else concerning Clough except that he was a resident of Boston by 1695; apparently he made, among other things, balances in a range of sizes for merchants and shopkeepers, marking his shop with a hand holding one of his products.[28] Occasionally over the course of

---

24 John Hammond, *The practical surveyor*, London, 1765, 4th ed. According to R. G. W. Anderson, J. Burnett and B. Gee, *Handlist of scientific instrument-makers' trade catalogues, 1600–1914*, Edinburgh, 1990, p. 39, a 16-page Heath and Wing catalog had appeared in 1757, a claim I have not been able to confirm. Presumably, these makers did not keep all the advertised instruments in stock, waiting for customers who walked into their shops; rather they would make to order.

25 Morrison-Low, op. cit. (n. 3), p. 290.

26 Samuel Clough, *Kalendarium Nov-Anglicanum, or An almanack for the year of our Lord 1707 ... calculated for the meridian of Boston*, Boston, 1707.

27 S. Clough, *Clough's farewell, an almanack for 1708*, Boston, 1707.

28 Jay Mack Holbrook, *Boston beginnings, 1630–1699*, Oxford, MA, 1980, p. 55. According to Marion Barber Stowell, "American almanacs and feuds", *Early American literature*

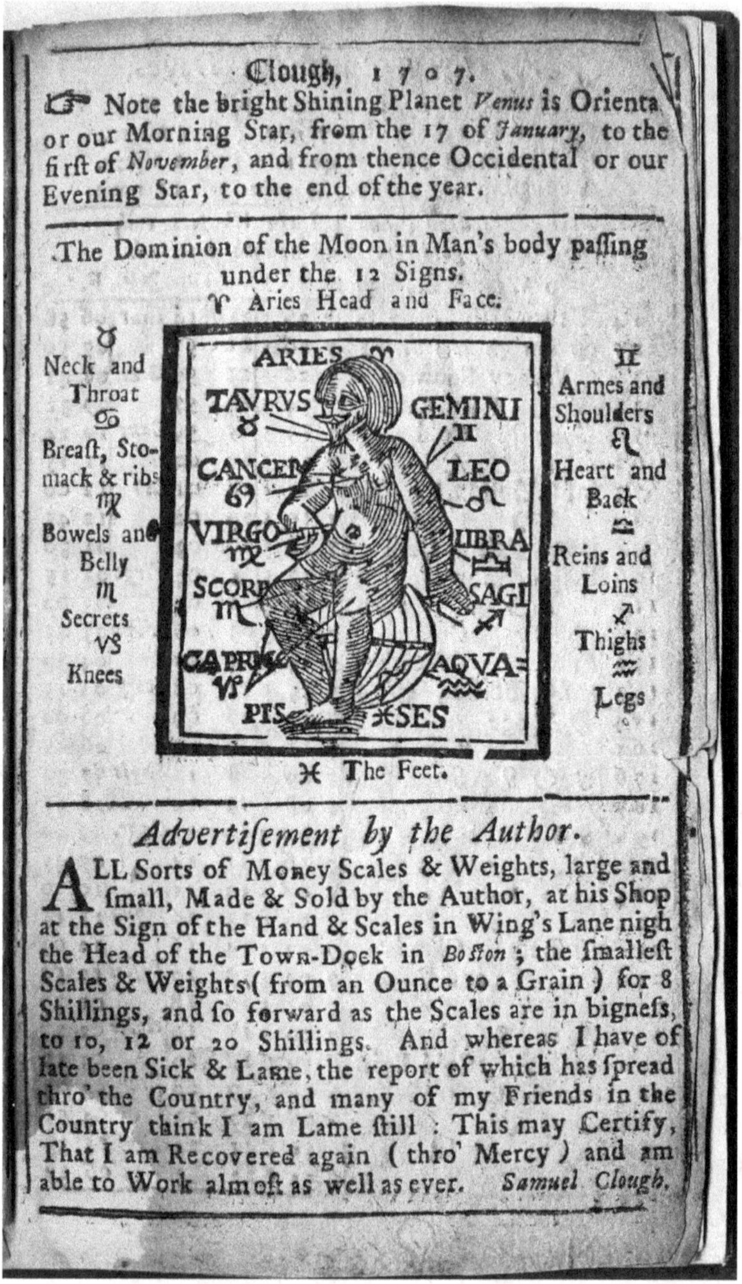

Clough, 1707.

☞ Note the bright Shining Planet *Venus* is Orienta or our Morning Star, from the 17 of *January*, to the first of *November*, and from thence Occidental or our Evening Star, to the end of the year.

The Dominion of the Moon in Man's body passing under the 12 Signs.

♈ Aries Head and Face.

♉ Neck and Throat
♋ Breast, Stomack & ribs
♍ Bowels and Belly
♏ Secrets
♑ Knees

ARIES ♈
TAVRVS ♉
GEMINI ♊
CANCER ♋
LEO ♌
VIRGO ♍
LIBRA ♎
SCORP ♏
SAGI ♐
CAPRIC ♑
AQVA ♒
PISSES

♊ Armes and Shoulders
♌ Heart and Back
♎ Reins and Loins
♐ Thighs
♒ Legs

♓ The Feet.

Advertisement by the Author.

ALL Sorts of Money Scales & Weights, large and small, Made & Sold by the Author, at his Shop at the Sign of the Hand & Scales in Wing's Lane nigh the Head of the Town-Dock in *Boston*; the smallest Scales & Weights (from an Ounce to a Grain) for 8 Shillings, and so forward as the Scales are in bigness, to 10, 12 or 20 Shillings. And whereas I have of late been Sick & Lame, the report of which has spread thro' the Country, and many of my Friends in the Country think I am Lame still: This may Certify, That I am Recovered again (thro' Mercy) and am able to Work almost as well as ever. *Samuel Clough.*

FIGURE 8.1 *Samuel Clough,* Almanack for 1707, *Boston, 1707, f.* [2r]. *The 'zodiac man,' providing instructions for astrologically-informed bloodletting, had graced printed almanacs since the fifteenth century and manuscripts since the twelfth century.*

AMERICAN ANTIQUARIAN SOCIETY, ALMANACS MASS C410, 1707.

the eighteenth century, advertisements for quadrants, compasses, Gunther's scales, spy glasses, scales and weights, carpenter's rules, etc. would appear in almanacs whose printers operated bookshops that sold mathematical instruments. Presumably, most of this apparatus was imported from Europe. But colonial almanacs usually advertised only books offered by their printers (cf. Benjamin Franklin).[29]

Most of the formal advertisements for mathematical instruments appeared in early American newspapers. My preliminary, and by no means complete, search of early American newspapers in the seven decades from 1730 through 1800 has located about 200 separate advertisements for mathematical instruments, mostly by makers but also by some retailers (a total of 87 names).[30] Our first conclusion, therefore, is that newspaper advertising for instruments was rare, sporadic and, as we shall see, concentrated among a few makers. As historians of advertising have emphasized, roughly 90% of the white population lived on farms in colonial America, producing most everything they needed or trading with neighbors. Little surplus cash circulated to which advertisements for consumer products could appeal. "The problem was rarely to find a market (and word of mouth took care of most of the exchange); the problem was finding someone with money to pay for it".[31] Our second conclusion, therefore, is that instrument makers did not target the general public with their advertising; their focus rather was on the practical mathematician or the practitioner who needed specialized apparatus to conduct his trade.

---

9 (1975), pp. 276–285, some colonial printers sold more than 50,000 almanacs a year.

29 Cf. Milton Drake, *Almanacs in the United States*, New York, 1962, 2 vols.: nos. 2198 (Baltimore, 1785), 5886 (New York, 1781), 13094, 13103, 13108 (Charleston, 1767, 1770, 1772).

30 In addition to J. Carter Harris, *The clock and watch makers American advertiser, being a compilation of advertisements of the trade, including those of the mathematical and philosophical instrument makers, published in the newspapers of the American colonies and states, from A.D. 1707 to 1800*, Wadhurst, 2003, I have conducted full-text searches in *America's historical newspapers, Early American newspapers*, Series 1–7, 1690–1922, www.readex.com/content/early-american-newspapers-1690-1922-series (accessed 15 March 2016). Currently containing more than 2,000 scanned newspaper runs, this proprietary database published by Readex continues to add titles and to extend runs but does not yet include all colonial newspapers, and full-text searching can be degraded by irregular fonts or linebreaks in the originals. Nonetheless, most major newspapers are included and my survey of ads is surely representative, if not exhaustive.

31 Charles Goodrum and Helen Dalrymple, *Advertising in America: The first 200 years*, New York, 1990, pp. 16–17.

Before turning to other trends, let me introduce the two most prolific of the eighteenth-century advertisers, both rather idiosyncratic not only in the frequency of their advertisements but also in their marketing strategies. As noted above, the earliest newspaper advertising instrument maker in America was Anthony Lamb of New York. Between December 1730 and March 1731, Lamb placed three nearly identically worded advertisements in Benjamin Franklin's *Pennsylvania Gazette*, announcing himself as a "mathematical instrument maker, from London" with "all sorts of instruments for sea or land" available in his New York shop at the "Compass and Quadrant".[32] Lamb's advertisements are noteworthy for several reasons. First, Franklin's weekly *Gazette* generally devoted one of its four pages to advertising, the overwhelming majority of which dealt with runaways (indentured servants, apprentices or slaves), real estate, or imported luxury goods. In the *Gazette's* first three years under Franklin's editorship (1729–1731), I find only four advertisements by craftsmen, two seeking raw materials (white horse hair for a wigmaker, hog's bristles for a brush maker), one offering services (bookbinder), and Lamb offering instruments he had made. Since most locally made items were undoubtedly consumed locally, colonial craftsmen at least in the 1730s rarely advertised. Lamb, on the other hand, advertised his New York shop in a Philadelphia newspaper. As Silvio Bedini has noted, Franklin's *Gazette* already by 1731 circulated widely through the American colonies, unlike the local *New York Gazette*, founded in 1725.[33] I find no other advertisements from New York merchants or craftsmen in the first three years of Franklin's *Gazette*. Lamb appears to be unconventional in aiming his advertising beyond local readers. Lamb's advertisement indicates that he worked in wood, ivory, brass or silver, suggesting that he considered his market to range from the poor rural surveyor to the wealthy sea captain. According to Franklin's account book, Lamb paid 3 shillings for each of these advertisements, a considerable amount, which might also explain why so few craftsmen purchased space in Franklin's newspaper.[34]

Lamb is also an outlier for the extended span of his advertising. After a fourteen-year hiatus, he would place fourteen more advertisements in

32 *Pennsylvania Gazette*, 3 and 15 December 1730, and 25 March 1731.

33 Bedini, "At the sign of the compass and quadrant...", op. cit. (n. 11), pp. 20 and 29.

34 Ibid., pp. 20–21. Franklin advertised a dozen almanacs for 3 shillings in the *Pennsylvania Gazette*, 2 October 1729. On 8 October 1736, the *Virginia Gazette* announced its advertising rates as 3 shillings per week, 2 shillings per subsequent week, "and as these papers will circulate as speadily as possible not only all over this, but also the neighbouring colonies, and will probably be read by some thousands of people [...] it is certainly the cheapest and most effectual method" of advertising.

New York newspapers from 1745 to 1775. His target clientele expanded beyond the practitioner; in 1745 he offered German flutes, drawing pens, protractors, parallel rulers, walking-stick spying-glasses, as well as billiard balls and heads of canes, and promised "ready money for old kettles or other brass".[35] In 1749, he offered the "late invented and most curious instrument call'd an octant",[36] perhaps in conjunction with Thomas Godfrey of Philadelphia who in 1731 had simultaneously invented with John Hadley in London the altitude-measuring device that would soon be called "Hadley's quadrant".[37] A 1753 Lamb advertisement refers to "Godfrey's new invented quadrant".[38] It is interesting that Lamb described as "newly invented" a device proposed twenty years earlier and advertised by the Boston maker, Joseph Halsy, Jr., already in 1738 as "Hadley's new invented quadrant or octant the best and exactest instrument for taking the latitude [...] as ever yet invented".[39] Bedini wondered whether Lamb waited until after Godfrey's death in 1749 to start making the new device or until after Hadley's 1734 patent had expired.[40] In any case, in 1760 Lamb advertised another new instrument, two graduated, concentric circles for computing the number of men in military regiments, invented, proclaims the ad, by Colonel William Haviland and "suitable for Gentlemen of the Army".[41] Haviland would become a significant British commander in the French and Indian War.

Lamb's final two advertisements appeared in 1779 and 1781 in New Haven newspapers. The seventy-seven year old maker still "makes and sells all sorts of instruments for land and sea" but also offered shaving boxes, tortoise shell rings, fans neatly folded, putty for mending China, hair pins and bodkins, maps, drawing books, pictures, and watch covers.[42] The once specialized mathematical instrument maker had become a retailer of luxury items, advertising no longer in New York but in the wealthy Connecticut town of New Haven. A similar focus appears in Lamb's only known printed broadside, a rococo

35 *New York Post-Boy*, 6 May 1745.

36 Ibid., 23 January 1749.

37 John Hadley, "Description of a new instrument for taking angles", *Philosophical transactions* 37 (1731), pp. 147–157.

38 *New York Mercury*, 11 June 1753.

39 *Boston Gazette*, 18 September 1738; Logan, op. cit. (n. 10), pp. 441–450.

40 Bedini, "At the sign of the compass and quadrant...", op. cit. (n. 11), p. 44. Cf. Warner, "Davis' quadrants...", op. cit. (n. 8), p. 39. The Boston maker John Dabney advertised Hadley's quadrants in the *Boston Evening Post*, 30 July 1739.

41 *New York Mercury*, 14 April 1760; Bedini, "At the sign of the compass and quadrant...", op. cit. (n. 11), pp. 53–63. For a later advertisement for this device, see the *New York Gazette*, 12 May 1777.

42 *Connecticut Journal*, 16 June 1779, 20 Dec 1781.

advertisement, c. 1755, engraved by a craftsman who had recently joined his shop (*Fig. 8.2*). The sheet features a likeness of Isaac Newton (Lamb's shop was now "At Sir Isaac Newton's Head") and nestles various mathematical instruments amidst the birds, foliage and scallops surrounding the text. Lamb's only illustrated advertisement, this broadside closely followed mid-century London designs; but it is the only such rococo ephemera I have found by an instrument maker in colonial America.[43] The other known eighteenth-century trade cards for American makers, very few in number, are quite plain (*Fig. 8.3*).

Our next most prolific advertiser is the otherwise little known Boston maker, Andrew Newell (1749–1798). Unlike any other maker in my sample, Newell launched what appears to have been a blitzkrieg of seventeen advertisements over twelve months in Boston newspapers; no other maker in my sample would buy so many advertisements over such a short time period. A closer look, however, reveals an advertiser ensnared in the politics of the American Revolution. In June and July 1773, Newell placed the identical advertisement five times in two Boston papers, the *Massachusetts Gazette and Post-Boy*, a staunchly Royalist paper, and the *Boston Evening Post* with a more impartial editorial stance. Newell defined himself as a mathematical instrument maker, noted that he had been apprenticed under the best masters in London, and indicated that he imports, makes, sells and mends "all kinds of mathematical instruments," naming none in particular.[44] Then in October to December 1773 he placed a different advertisement eight times in the *Evening Post* and the *Massachusetts Spy*, among the most radical newspapers defending the patriot cause in the American colonies, published by Isaiah Thomas. The new advertisement lists dozens of individual instruments for drawing, surveying and navigating, plus thermometers, "Dollon's patent telescopes", spectacles, and a large array of stationary goods.[45] Newell's final advertisement appeared four times in May to June 1774 in the *Evening Post* and the *Boston Gazette*, the latter an even more radical paper edited by Benjamin Edes, whom the British called the "trumpeter of sedition." This short advertisement merely reports

---

43 The design of Lamb's broadside is very similar to that printed for the London maker, John Gilbert, undated, but probably also 1750s, reproduced in Calvert, op. cit. (n. 22), pl. 24. Cf. Clifton, op. cit. (n. 3), p. 112. The only other eighteenth-century American broadside I have found is by the Philadelphia makers Benjamin Condy & Son, c. 1784, an unadorned text listing their "mathematical & optical instruments made and sold": Library Company of Philadelphia, Am 1784 Ben Cond 9068.F (21 × 17 cm).

44 *Massachusetts Gazette and Boston Post-Boy*, 21 June, 5 July, 12 July 1773; *Boston Evening Post*, 21 June, 12 July 1773.

45 *Boston Evening Post*, 11 October, 18 October, 25 October 1773; *Massachusetts Spy*, 14 and 21 October, 26 November, 2 December, 16 December 1773.

FIGURE 8.2 *Anthony Lamb, broadside, c. 1755.*
COLLECTION OF THE NEW-YORK HISTORICAL SOCIETY, BROADSIDE SY1755 NO. 8, 32 × 22 CM.

FIGURE 8.3 *Samuel Thaxter, trade card, c. 1792.*
AMERICAN ANTIQUARIAN SOCIETY, EPHEMERA TRADE 0675, 11 × 8 CM.

that Newell had just imported "all sorts of mathematical instruments" (listing none in particular) and that "the present prospects of affairs in the town induces him to sell cheap for cash only." I have found no further Newell advertisements; he is listed in the first Boston city directory of 1789.[46] In any case, we might speculate that Newell, recently arrived from London, strategically spread his advertising to adapt to an exceedingly politicized Boston market in the mid-1770s. His multiple advertisements certainly would have saturated Boston's newspaper-reading public.[47]

Looking more comprehensively at my sample of the 200 eighteenth-century newspaper advertisements we find little change in the typographic layout and

46 *Boston Evening Post*, 23 May, 6 June 1774; *Boston Gazette*, 23 May, 30 May 1774. Cf. Bedini, *Early American scientific instruments…*, op. cit. (n. 9), pp. 106–107; Timothy M. Barnes, "Loyalist newspapers of the American revolution, 1763–1783: A bibliography", *Proceedings of the American Antiquarian Society* 83 (1974), pp. 217–240; Carol Sue Humphrey, *This popular engine: New England newspapers during the American Revolution, 1775–1789*, Newark, 1992. Newell must have imported most, if not all, of what his advertisements offer; surely he did not make thermometers or Dollond telescopes.

47 According to Heyd, op. cit. (n. 2), p. 17, in 1750 the average circulation of Boston newspapers was 600 sheets; by 1775, the wildly popular *Massachusetts Spy* attained a circulation of 3,500 sheets. And most of these sheets had multiple readers.

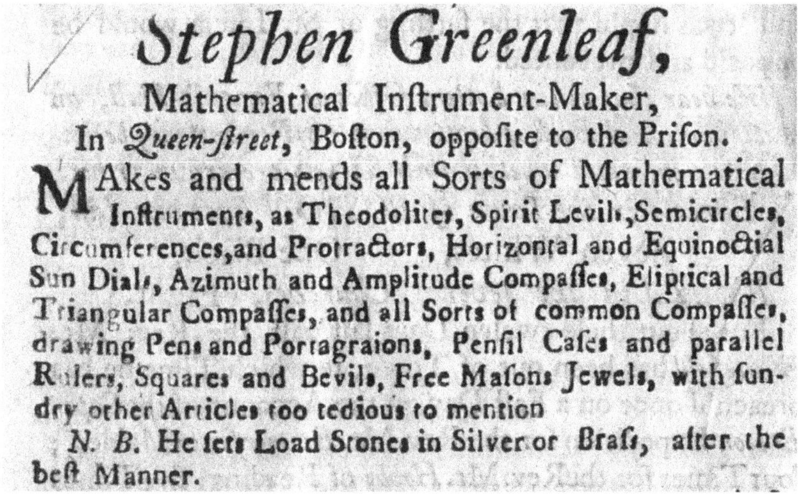
Stephen Greenleaf,
Mathematical Inſtrument-Maker,
In *Queen-ſtreet*, Boſton, oppoſite to the Priſon.
MAkes and mends all Sorts of Mathematical Inſtruments, as Theodolites, Spirit Levils, Semicircles, Circumferences, and Protractors, Horizontal and Equinoctial Sun Dials, Azimuth and Amplitude Compaſſes, Eliptical and Triangular Compaſſes, and all Sorts of common Compaſſes, drawing Pens and Portagraions, Penſil Caſes and parallel Rulers, Squares and Bevils, Free Maſons Jewels, with ſundry other Articles too tedious to mention
N. B. He ſets Load Stones in Silver or Braſs, after the beſt Manner.

FIGURE 8.4 *Advertisement format listing individual items.* Boston Gazette, *18 June 1745, p. 4.* AMERICAN ANTIQUARIAN SOCIETY, NEWSD MA BOST GAZE.

content of the advertisements between 1730 and 1800. They generally appear on the last page of the newspaper, placed in an advertising section separated from the other content of the issue. Most advertisements present the subscriber's name, identify him as a "mathematical instrument maker", locate the shop by local landmarks or street names, and provide either long lists of individual items being offered (often ending with "sundry other articles too tedious to mention", see *Fig. 8.4*) or shorter lists of types of items (e.g., surveying compasses, telescopes "for land and sea"). Particular prices are almost never named, but many advertisements promise that service will be "prompt", workmanship "neat", and prices "reasonable". Only in the 1770s do small images start to accompany some of the advertisements, usually of a figure holding Hadley's quadrant, undoubtedly copied from a common carved, painted wooden sign announcing on the street a mathematical maker's shop (see *Fig. 8.5*).[48] A large majority of the makers "make and mend" or "make, sell and mend" instruments; indeed, we might guess that many spent as much, if not more time, completing repairs than making new items for their customers.

48 The Boston maker William Williams had since 1768 advertised his shop at the "Sign of Admiral Vernon". In 1771, his ads begin to include a small woodcut of an admiral sighting with a Hadley's quadrant, surely referencing the wooden shop sign. See *Boston Gazette*, 17 October 1768; *Massachusetts Spy*, 15 August 1771.

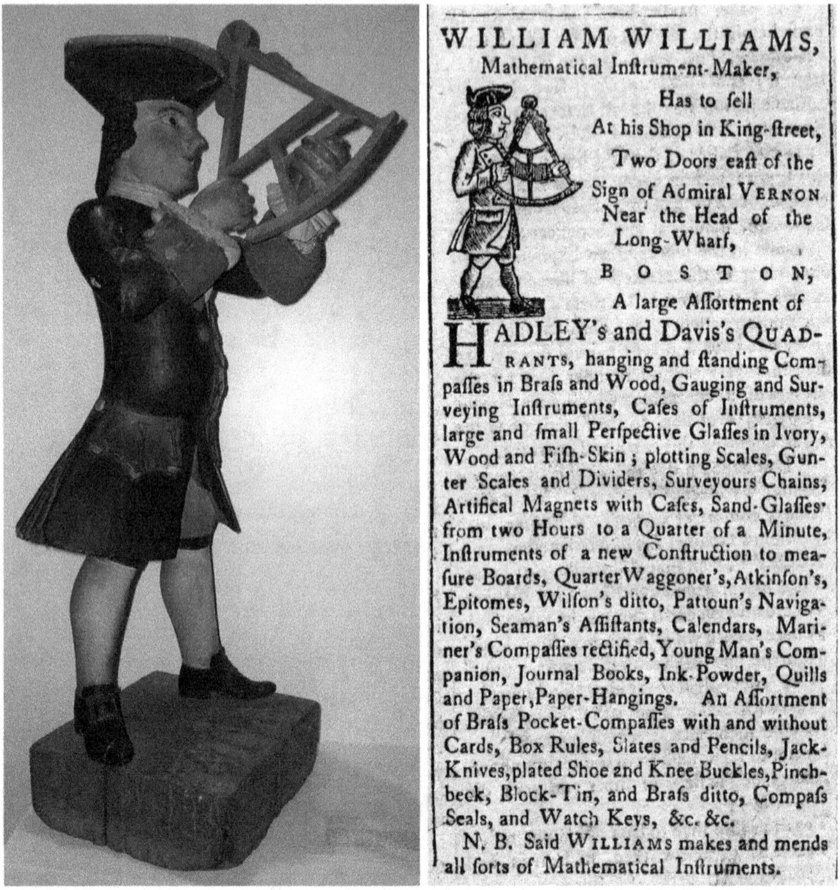

FIGURE 8.5 *Standing admiral with Hadley's quadrant as an advertising icon for mathematical instruments. Clark Elliott's shop figure, New London, Connecticut, c. 1770, carved wood, 90 cm high (New London County Historical Society) and William Williams' advertisement in the* Massachusetts Spy, *15 August 1771, p. 96.*
AMERICAN ANTIQUARIAN SOCIETY, NEWSD MA BOST MASS.

Very few makers advertised over extended periods of time. Rather, they would advertise when they moved their shop to a new location in the town or (far more rarely) moved from one town to another. They would advertise when they had something newly invented to offer. A 1771 advertisement offers "sliding rules of a new construction";[49] another from the same year mentions "instruments

49 James Condy in the *Massachusetts Spy*, 25 July 1771.

of a new construction to measure boards".[50] After the founding of the new nation, a maker in 1790 proudly christened his invention as the "American theodolite".[51] An advertisement from 1794 promises quadrants "with the latest improvements";[52] another maker from that year enigmatically referred to "several new philosophical instruments of his own invention".[53] In 1768, one of the colonies' most renowned makers, David Rittenhouse, published a lengthy description of a large orrery that he had designed and was building, emphasizing its novelty. But the text appears in the newspaper as a news column rather than an advertisement and does not explicitly advertise any products by Rittenhouse.[54] Indeed, Rittenhouse apparently never advertised formally in any of Pennsylvania's newspapers.

Over time, several changes do appear in my sample. Starting in the 1760s, retailers increasingly began to advertise mathematical instruments as one of the many types of goods available in their shops. Many retailers were booksellers and stationers; other merchants sold instruments along with luxury household items, glassware and china, pewter, musical instruments, fancy fabrics, medicines, tropical fruit, coffee, wine and candy. Retailers often advertised the arrival of large shipments from London. Judging from the advertisements, we might guess that as many mathematical instruments were imported and sold by retailers as were made and offered by local makers in their own shops, especially in the middle third of the century.

Another broadening of the market appears in the self-designations of the advertisers. Over the century, most of the advertisements refer to "mathematical instruments". In 1761, announcing his move from New York to Philadelphia, the maker James Ham advertised a long list of "mathematical and optical instruments" in his shop, the first time in my sample that I have found these two adjectives together.[55] In 1796, William Richardson in Philadelphia advertised himself as an "optician" who sold and repaired both mathematical and optical items.[56] Not until 1784 would Robert Clark, in Boston, describe himself as a "mathematical, optical and philosophical instrument maker", and offer, for the first time as far as I know, an electric machine and mechanical powers, both

50 William Williams in the *Massachusetts Spy*, 15 August 1771.

51 Robert Coram in the *Delaware Gazette*, 2 August 1790.

52 Isaac Greenwood in the *Providence Gazette*, 21 June 1794.

53 Joseph Gatty in the *New York Diary or Evening Register*, 22 October 1794.

54 *Pennsylvania Chronicle*, 16 May 1768. Cf. Francois D. Bud Uzes, "David Rittenhouse: Colonial surveyor and instrument-maker", *Rittenhouse* 5 (1991), pp. 1–16.

55 *Pennsylvania Gazette*, 19 March 1761.

56 *Finlay's American naval and commercial register* (Philadelphia), 12 February 1796.

philosophical instruments, along with the usual quadrants, compasses, and Gunter's scales.[57] In 1792, after returning from five years of military service with the patriot army, the Philadelphia maker Thomas Biggs advertised "a large and elegant assortment of mathematical, philosophical and optical instruments, of almost every description".[58] Yet Clark and Biggs remained outliers; in 1793, an anonymous "Philosophicus" complained in a Boston newspaper about the lack of philosophical makers in the new nation. Mathematical makers could "barely furnish a few second-hand instruments, [...] [and] none of the nobler sorts of philosophical instruments are to be procured anywhere in the country. Every gentlemen, wishing to indulge the love of philosophy, must be at the expense of sending to Europe for the whole of his apparatus".[59] Philosophicus would have to wait. Not until the 1826 would an English mechanic move to Boston and begin making "apparatus for illustrating the various sciences", items used in his own public lectures. In 1834, this emigrant, Timothy Claxton (b. 1790), would publish the first multi-paged *Catalogue of philosophical apparatus* in America, listing with prices items for demonstrating mechanics, hydrostatics, pneumatics, chemistry, optics and aimed at school teachers and lecturers in mechanics' institutes rather than at mathematical practitioners (*Fig. 8.6*).[60]

Occasionally the advertisers explicitly described their intended customers. In 1756, the New London maker Clark Elliott advertised instruments "for any gentlemen, sailors, surveyors, guagers [sic]".[61] James Ham, in the 1761 advertisement noted above that offered both mathematical and optical instruments, described his shop as a place "where Masters of Ships and others may be faithfully served".[62] Garrat Noel, a New York retailer, offered in 1763 sundials "for

57 *Maryland journal and the Baltimore advertiser*, 9 April 1784.

58 *Federal Gazette* (Philadelphia), 19 September 1792.

59 *Columbian Centinel* (Boston), 6 April 1793, p. 2.

60 Timothy Claxton, *Memoir of a mechanic: Being a sketch of the life of Timothy Claxton, written by himself, together with miscellaneous papers*, Boston, 1839, p. 65; Timothy Claxton and Joseph M. Wightman, *Claxton's catalogue of philosophical apparatus for sale at 23 Water Street, Boston*, Boston, 1834; T. Claxton, *Catalogue of philosophical apparatus manufactured and sold by Claxton & Wightman, No. 33 Cornhill, Boston*, [Boston], 1835. Claxton's catalogs are uniquely preserved at the Beinecke Library, Yale University, Scientific Miscellany 11. For Claxton's intended clientele, see William Channing Woodbridge, "Claxton's philosophical apparatus", *American annals of education and instruction* 5 (1835), p. 479. To support my claim that Claxton's are the first multi-page catalogs with prices printed in America, cf. Anderson, Burnett, and Gee, op. cit. (n. 24), *passim*.

61 *Connecticut Gazette*, 3 July 1756.

62 *Pennsylvania Gazette*, 19 March 1761.

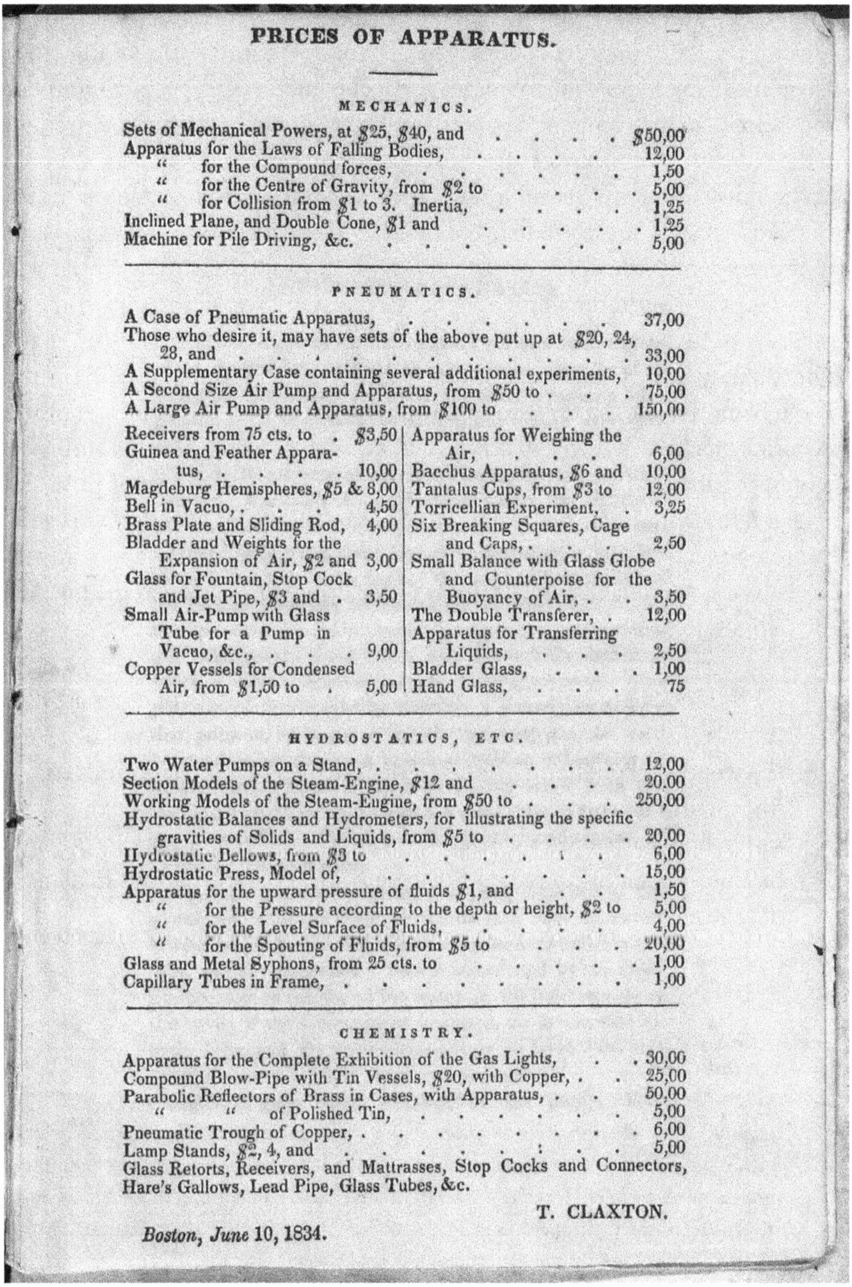

PRICES OF APPARATUS.

MECHANICS.

Sets of Mechanical Powers, at $25, $40, and	$50,00
Apparatus for the Laws of Falling Bodies, ...	12,00
" for the Compound forces, ...	1,50
" for the Centre of Gravity, from $2 to	5,00
" for Collision from $1 to 3. Inertia, ...	1,25
Inclined Plane, and Double Cone, $1 and	1,25
Machine for Pile Driving, &c.	5,00

PNEUMATICS.

A Case of Pneumatic Apparatus, ...	37,00
Those who desire it, may have sets of the above put up at $20, 24, 28, and	33,00
A Supplementary Case containing several additional experiments,	10,00
A Second Size Air Pump and Apparatus, from $50 to	75,00
A Large Air Pump and Apparatus, from $100 to	150,00

Receivers from 75 cts. to .	$3,50	Apparatus for Weighing the Air, ...	6,00
Guinea and Feather Apparatus, ...	10,00	Bacchus Apparatus, $6 and	10,00
Magdeburg Hemispheres, $5 &	8,00	Tantalus Cups, from $3 to	12,00
Bell in Vacuo, ...	4,50	Torricellian Experiment, .	3,25
Brass Plate and Sliding Rod,	4,00	Six Breaking Squares, Cage and Caps, ...	2,50
Bladder and Weights for the Expansion of Air, $2 and	3,00	Small Balance with Glass Globe and Counterpoise for the Buoyancy of Air, . .	3,50
Glass for Fountain, Stop Cock and Jet Pipe, $3 and .	3,50	The Double Transferer, .	12,00
Small Air-Pump with Glass Tube for a Pump in Vacuo, &c., ...	9,00	Apparatus for Transferring Liquids, ...	2,50
Copper Vessels for Condensed Air, from $1,50 to .	5,00	Bladder Glass, ...	1,00
		Hand Glass, ...	75

HYDROSTATICS, ETC.

Two Water Pumps on a Stand, ...	12,00
Section Models of the Steam-Engine, $12 and	20.00
Working Models of the Steam-Engine, from $50 to	250,00
Hydrostatic Balances and Hydrometers, for illustrating the specific gravities of Solids and Liquids, from $5 to	20,00
Hydrostatic Bellows, from $3 to	6,00
Hydrostatic Press, Model of, ...	15,00
Apparatus for the upward pressure of fluids $1, and . . .	1,50
" for the Pressure according to the depth or height, $2 to	5,00
" for the Level Surface of Fluids, ...	4,00
" for the Spouting of Fluids, from $5 to	20,00
Glass and Metal Syphons, from 25 cts. to	1,00
Capillary Tubes in Frame, ...	1,00

CHEMISTRY.

Apparatus for the Complete Exhibition of the Gas Lights, . .	30,00
Compound Blow-Pipe with Tin Vessels, $20, with Copper, . .	25,00
Parabolic Reflectors of Brass in Cases, with Apparatus, . .	50,00
" " of Polished Tin, ...	5,00
Pneumatic Trough of Copper, ...	6,00
Lamp Stands, $2, 4, and	5,00

Glass Retorts, Receivers, and Mattrasses, Stop Cocks and Connectors, Hare's Gallows, Lead Pipe, Glass Tubes, &c.

T. CLAXTON.

Boston, June 10, 1834.

FIGURE 8.6 *The first instrument catalogue printed in America.* Claxton's catalogue of philosophical apparatus for sale at 23 Water Street, Boston, *Boston, 1834, p.* [2]. BEINECKE RARE BOOK AND MANUSCRIPT LIBRARY, YALE UNIVERSITY, SCIENTIFIC MISCELLANY 11.

gentlemen's gardens".[63] In 1791 Thomas Biggs in New York proposed special rates for the "ship masters of New York".[64] But the vast majority of the advertisements in my sample do not target particular audiences; simply announcing "mathematical instruments" was apparently adequate to situate the products in the commercial culture of eighteenth-century America.[65]

The advertisements reveal a few other isolated examples of how makers sought to market their products. Unlike eighteenth-century London, where prominent makers such as George Adams or Benjamin Martin delivered public lectures and authored popular books to boost interest in their instruments, only two of my advertisers employed such tactics. In 1743, the Boston maker John Dabney advertised "on Monday, Tuesday and Thursday evenings, from five to eight o'clock, for the entertainment of the curious, the magic lanthorn an optick machine, which exhibits a great number of wonderful and surprising figures, prodigious large, and vivid, at half a crown each, old tenor [2 ½ shillings, a not inconsequential amount]".[66] In the fall of 1747, the *Boston Evening Post* announced that the Newport, Rhode Island printer, clock- and compass-maker, William Claggett (1696–1749) was performing in Boston at the house of one Captain John Williams:

> for the entertainment of the curious [...] a great variety of curious experiments of the most surprizing effects of electricity, wherein will be shown the wonderful phenomena of electrical attraction, repulsion and flamisic force; particularly the new method of electerising several persons at the same time, so that fire shall dart from all parts of their bodies, as the same has lately been exhibited, to the astonishment of the curious in all parts of Europe.[67]

63 *New York Gazette*, 13 June 1763.

64 *Daily Advertiser* (New York), 11 April 1791.

65 A similar study of advertising by shopkeepers and tradesmen in the American colonies finds repeated references to customers as "gentlemen and ladies", rhetoric pointing to shops as places where everyone joined the gentry class. I do not find this language in the advertising of the instrument makers. See Richard L. Bushman, "Shopping and advertising in colonial America", in Cary Carson, Ronald Hoffmann, and Peter J. Albert (eds.), *Of consuming interests: The style of life in the eighteenth century*, Charlottesville, 1994, pp. 233–251.

66 *Boston Evening Post*, 19 December 1743.

67 *Boston Evening Post*, 24, 31 August and 7 September 1747. Already in 1746 Claggett had been experimenting with electricity in Newport. See *Pennsylvania Gazette*, 27 January 1747.

For ten shillings (old tenor), people could see Claggett's experiments "at any time of the day". A later account claimed that Claggett, during three weeks of demonstrations in Boston, earned what would have been the extraordinary sum of £ 1,500, which he was said to have donated to charity![68]

Often considered among the earliest public electrical amusements conducted in America, Claggett's demonstrations were continued into November by Williams. In October of 1747, two Boston newspapers announced that "the wonderful and surprising operations in electricity lately shewn by Mr. Clagget[t] in Boston, are now performed by Daniel King", a mathematical instrument maker in Salem, prompting Williams angrily to advertise that he had the original electrical machine made by Claggett. Apparently King had secured (or made) another machine. In any case, eastern Massachusetts seemed awash in electrical entertainment during the fall of 1747.[69]

No evidence suggests, however, that Claggett or King sought to turn their public performances into increased sales of instruments. None of these men advertised further public performances and the phenomenon of instrument makers as lecturers or entertainers seems not to have spread in eighteenth-century America. Indeed, the vast majority of the advertisements in my sample are written in rather sedate prose; hyperbole, hucksterism, and hype do not characterize the genre... although the repeated emphasis on "reasonable" prices might suggest a desire to appear sober in a wild world of eighteenth-century advertising.

Only a few of the makers mention shop signs, presumably carved wooden figures, in their newspaper advertising. Of the 87 shops that advertised in my sample, only eleven directed readers toward a sign. The advertisements refer to the "Compass & Quadrant", "Davis Quadrant", "Hadley's Quadrant",

---

68 Arthur Ross, *A discourse, embracing the civil and religious history of Rhode-Island*, Providence, 1838, pp. 35–36.

69 Williams in *Boston Evening Post*, 28 September, 5 and 19 October 1747, 2 November 1747; *Boston Gazette*, 6 October 1747; Daniel King in ibid., 13 October 1747; *Boston Evening Post*, 12, 19 and 26 October 1747. For King's widow's attempt to sell Claggett's electrical machine after her husband's death, see *Boston Weekly Post-Boy*, 20 and 27 November 1749. For more on this interesting episode, see William Northrop Morse, "Lectures on electricity in colonial times", *New England quarterly* 7 (1934), pp. 364–374; I. Bernard Cohen (ed), *Benjamin Franklin's experiments... with a critical and historical introduction*, Cambridge, MA, 1941, pp. 53–56; Richard L. Champlin, "High time: William Claggett and his clockmaking family", *Newport history: Bulletin of the Newport Historical Society* 47, no. 3 (1974), pp. 157–185. In 1743 Adam Spencer, an itinerant Scottish lecturer, had in Boston performed electrical demonstrations as part of his course of public lectures on experimental philosophy. See *Boston Weekly Post-Boy*, 30 May 1743.

"Hadley's Compass and Quadrant Card", "Dial", "Globe", "Hand & Scales", and "Looking Glass", all quite literal representations of mathematical instruments. The more vague "Seven Stars" might have evoked astronomy or navigation. Only three individuals appeared: "Franklin's Head", "Sir Isaac Newton", and "Admiral Vernon", presumably Edward Vernon, a popular British admiral in the 1730–1740s, after whom George Washington's brother, who had served under the admiral, named his estate.[70] The latter figure, depicted in military uniform, taking a sighting with Hadley's quadrant, was quite popular for sellers of navigational instruments; numerous exemplars remain extant in American and British collections (*Fig. 8.5*); the figure even appears in stories by Charles Dickens and Nathaniel Hawthorne.[71]

Several of the advertisers included autobiographical details, presumably to polish their image. Perhaps a dozen or more indicated that they were "from London". Only one emigrant in my sample indicated that he came from outside Britain, Joseph Gatty "artist [a term commonly used for French instrument makers in the late eighteenth and early nineteenth centuries, but not carried over to North America] from Italy" in 1794, a glass blower who offered "several new philosophical instruments of his own invention".[72] William Hinton in 1772 wrote that "as he is a young beginner, he flatters himself, he shall meet with encouragement".[73] Others touted their training, not their youth. Enos Doolittle's advertisement, also in 1772, noted that he had been apprenticed "with the most

70 Lamb in *Pennsylvania Gazette*, 3 December 1730; Walpole in *Pennsylvania Gazette*, 14 August 1740; Ham in *Pennsylvania Gazette*, 19 March 1761; Dorsey in *Norfolk Herald*, 15 April 1795; Heisely in *Maryland Chronicle*, 18 January 1786; Newell in *Boston Gazette*, 23 May 1774; Clough, *Kalendarium Nov-Anglicanum…*, op. cit. (n. 26); Duyckinck in *New York Mercury*, 19 July 1762; Denegan in *Pennsylvania Evening Herald*, 17 March 1787; Richardson in *Finlay's American Naval and Commercial Register* (Philadelphia), 12 February 1796; Williams in *Massachusetts Spy*, 14 July 1773; Williams in *Boston Gazette*, 17 October 1768. The story of the Admiral Vernon sign, apparently commissioned by William Williams in Boston in 1768, subsequently used by Andrew Newell and then by William's successor, Samuel Thaxter, and now preserved in the Peabody Museum, has been nicely told by Philip C. F. Smith, "The little men: Carved shop signs of the navigating instrument sellers", *American Neptune* 34 (1974), pp. 197–210. For a photograph of this sign, displayed at No. 1 Long Wharf in Boston for nearly 150 years, see Bedini, *Early American scientific instruments…*, op. cit. (n. 9), p. 77; "Report of the committee on the rooms", *Proceedings of the Bostonian Society* 36 (1917), pp. 14–16.

71 Smith, op. cit. (n. 70), pp. 197–210; Erwin O. Christensen, *Early American wood carving*, Cleveland, 1952, pp. 37–64.

72 *Diary or Evening Register* (New York), 22 October 1794.

73 *New York Gazette*, 18 May 1772.

noted workman in this colony".[74] John Gould, newly arrived from London in 1794, went further: "having done business for persons of the most approved scientific knowledge in the various departments in London, and has bro't into this country tools and mechanics suitable, solicits the encouragement of the public".[75] This is the only time the term "scientific" appears in my sample of eighteenth-century makers' advertisements!

Unlike Gould, most American makers and importers did not reference competition between British and American products in their advertisements. I have found only scattered examples of a retailer naming a London maker of the instruments he had imported. In the middle of the Revolutionary War, the printer of a royalist newspaper in Williamsburg offered a set of surveying instruments "made by Heath & Wing, London, of the latest construction".[76] Otherwise, British 'brands' are not emphasized (but see the Dollond patent telescopes mentioned above by Newell). American makers rarely compared theirs to European products. It is hence surprising to see Isaac Greenwood, a longtime instrument maker and "ivory turner" first in Boston and by 1786 in Providence, assure readers in 1788 that his articles "shall be made and repaired equal to any in Europe, and as cheap (for the same pay) as can be done on this continent. If the observation was more generally believed, the encouragement of our own manufactures will have as great a tendency to promote our national welfare as that of establishing a Federal Government, he should not doubt of meeting with success".[77] American advertisers also did not generally reference local competition. An exception was Thomas Biggs who in 1791 offered "time glasses, telescopes and other nautical instruments at the ship chandlery prices, in three months credit; [adding that] his instruments will be found to be of a superior quality to those in general furnished from the ship chandler, and being offered at the same prices, and terms of payment, he hopes will be a stimulus to excite the patronage of the merchants and ship masters of New York".[78] Biggs, the maker, is here directly challenging the retailing chandlers; but I find no other examples of such marketing strategies in my sample.

The advertisements provide some hints about the circulation of instruments in eighteenth-century America, about Greeley's "proper scope". Above, we

74 *Connecticut Courant*, 15 December 1772.

75 *Dunlap and Claypoole's American Daily Advertiser* (Philadelphia), 4 July 1794. See also Clifton, op. cit. (n. 3), p. 116.

76 John Dixon in *Virginia Gazette*, 11 July 1777.

77 *United States Chronicle* (Providence), 10 July 1788.

78 *Daily Advertiser* (New York), 11 April 1791.

noted that the earliest known newspaper advertisement, purchased by Lamb in 1730, appeared in a Philadelphia paper even though Lamb's shop was in New York and that he later advertised in a New Haven newspaper. I have found only one other example of such long-distance advertising. In 1773, a retailer in Petersburg, Virginia (population 2,800 in 1790) advertised the arrival of a shipment of surveying instruments from London in the *Virginia Gazette*, published in Williamsburg (population 1,300 in 1790), about 50 miles to the east.[79] Every other advertisement in my sample places the advertiser in the town or city in which its newspaper was published. Eighteenth-century newspaper advertising was local. Likewise, I have found only one example of a maker touting the wide circulation of his instruments. In 1795, the Fredericktown, Maryland maker, Frederick Heisely, moved to Lancaster, Pennsylvania to join forces with a clockmaker there. Their advertisement in the *Lancaster Journal* introduced Heisely as having made instruments for ten years, claiming that "his instruments are scattered from one extremity of the state of Maryland to the other, as well as through Virginia and Kentucky, and are universally approved of".[80] Only one advertisement in my sample, in 1796, explicitly solicited non-local customers: "orders from any part of the U.S. will be punctually attended to".[81] Although we do not know how widely the issues of any particular newspaper circulated, the rhetoric of the eighteenth-century advertisements, at least for mathematical instruments, remained local, focused on the town in which the newspaper was published.

Finally, we must note several extraordinary advertisements that appeared in the early 1750s in the *New York Gazette*, extraordinary because they were placed by a woman, the only female maker in my sample, and because they state that she "grinds all sorts of Optick Glasses to the greatest Perfection". As historian Huib J. Zuidervaart recently has discussed, one Balthasar Sommer and his wife in the 1720s had opened an optical shop in Amsterdam, grinding glass and offering microscopes, telescopes, and optical glasses. Their products entered many collections assembled by wealthy Amsterdam merchants. In 1733, Balthasar died and his widow continued the workshop; in 1749 she emigrated to New York and soon advertised "microscope glasses, spying-glasses of all lengths, spectacles, reading-glasses [...] all at the most reasonable

79 John Thomson in *Virginia Gazette* (Williamsburg), 10 June 1773.

80 *Lancaster Journal*, 22 January 1796. See also S. A. Bedini, "Frederick Heisely: Clock and compass maker", *Rittenhouse* 3 (1989), pp. 120–124.

81 *Maryland Journal and Baltimore Advertiser*, 12 May 1796.

rates".[82] These advertisements provide the earliest known evidence of an optician working glass in colonial America. Yet despite her advertising, widow Sommer's New York shop disappeared without a trace after 1753. The earliest documented spectacles made in America apparently date from the 1790s.[83]

Conclusion and Epilogue

If my sample is representative of the advertising purchased by eighteenth-century American instrument makers, we might conclude that even though the textual content of American advertisements essentially mirrors that of British advertisements from the same period, significant differences emerge between the advertising strategies in the two contexts. Americans advertised less frequently than their English counterparts, used trade cards or broadsides more rarely, and turned to the multi-page printed trade catalog about 80 years later. Americans did not employ public performances as part of their advertising. American makers tended to congregate in the larger towns and cities, and their advertisements likewise appeared more frequently in newspapers printed in those urban centers (patterns presumably found also in Britain). And the demand for philosophical instruments, by "gentlemen" with non-economic interests in natural philosophy or science, arose later in America than in Britain. Not surprisingly, therefore, it would be mathematical instruments—apparatus for surveying, navigating, measuring and quantifying—that would dominate the market in eighteenth-century America.

Turning again to Greeley's "proper scope", can we evaluate the impact of the makers' advertising strategies? Were their sales mostly local? How widely did their instruments circulate? Eighteenth-century account books, kept by

82 *New York Gazette*, 28 October and 18 November 1751, 21 May 1753; Huib J. Zuidervaart, "The 'invisible technician' made visible: Telescope making in the seventeenth and early eighteenth-century Dutch Republic", in A. D. Morrison-Low et al. (eds.), *From earth-bound to satellite: Telescopes, skills and networks*, Leiden, 2012, pp. 41–102: 95–96.

83 S. A. Bedini, *Thinkers and tinkers: Early American men of science*, New York, 1975, pp. 215–216; Rita Reif, "Fashion in glasses, sights to sore eyes", *New York Times*, 21 December 1997, p. AR38; Sara J. Schechner, "How telescopes came to New England, 1620–1740", in Marcus Granato and Marta C. Lourenço (eds.), *Scientific instruments in the history of science: Studies in transfer, use and perservation*, Rio de Janeiro, 2014, pp. 69–78; S. J. Schechner, "Telescopes in colonial and federal America, 1620–1820", (forthcoming).

instrument makers, are exceedingly rare and require more scrutiny than I can give here. But I have found several case studies that may prove useful.[84]

Thomas Greenough (1710–1785) was a Boston selectman (a member of the town's governing body in the eighteenth century), merchant, real estate magnate, importer, instrument maker, and part owner of a plantation in Antigua who actively supported the Revolution. His family papers include invoices and shipping lists (no account books, per se) from the 1740–1750s. He received bulk shipments of instruments (compasses, hour glasses, telescopes, quadrants, Gunter's scales) and books (e.g., James Atkinson's *Epitome of the whole art of navigation* and Archibald Patoun's *Complete treatise on practical navigation*) from London stationers (see *Fig. 8.7*) and ships' captains, and sold the same to Boston retailers. He purchased 72 dozen (!) compass glasses from Edmund Quincy and Sons, another Boston merchant. He repaired many compasses for various individuals, including Samuel Grant, Boston selectman, and John Rowe, another selectman, merchant, property developer, slave dealer and smuggler (Rowe's tea filled Boston's harbor after the Tea Party). Unfortunately, the documents do not indicate which items Greenough made and which he merely circulated as a trader. But generally, it would appear as if most of Greenough's customers lived in Boston.[85]

Clark Elliott (1732–93) was an instrument maker in New London, Connecticut (like Boston a major shipping port in the colonies) who had been apprenticed to Thomas Greenough. His account book for 1771–1774 and 1782–1793 records hundreds in transactions with individual customers (*Fig. 8.8*). "Sold an old quadt of Joseph Swan [...] sold one Hadley's quadt for Capt. Tho. Fanning of Norwith [...] Wait Rathborn took Barzilei Beebes octant".[86] Prices are rarely mentioned. Many of Elliott's sales were on consignment with both parties' names recorded (337 separate names appear in the accounts, of which only thirty-three appear more than once). His accounts suggest that his business peaked in 1782–1785 and that he sold three times as many items on consignment as new (*Table 4*). The account book also lists 503 wooden Davis quadrants

84 I have not yet examined the account books of William Hart (1734–1812), an instrument maker in Portsmouth, New Hampshire, preserved at the New Hampshire Historical Society, 1984-040 and 1984-073 (A).

85 Cf. David S. Greenough papers, Massachusetts Historical Society, Boston, Ms. N-355, boxes 4–6. Note that Mount & Page, a London stationer's firm since the 1710s, had printed many editions of Atkinson's *Epitome* since 1724, and would print many editions of Patoun's *Complete treatise*, starting in 1751. See Peter Barfoot, *Universal British directory of trade and commerce*, 3 vols., London, 1790–1794, v. 1, p. 235.

86 New London County Historical Society, Clark Elliott shop records, MsBd 158, ff. 1r-2r.

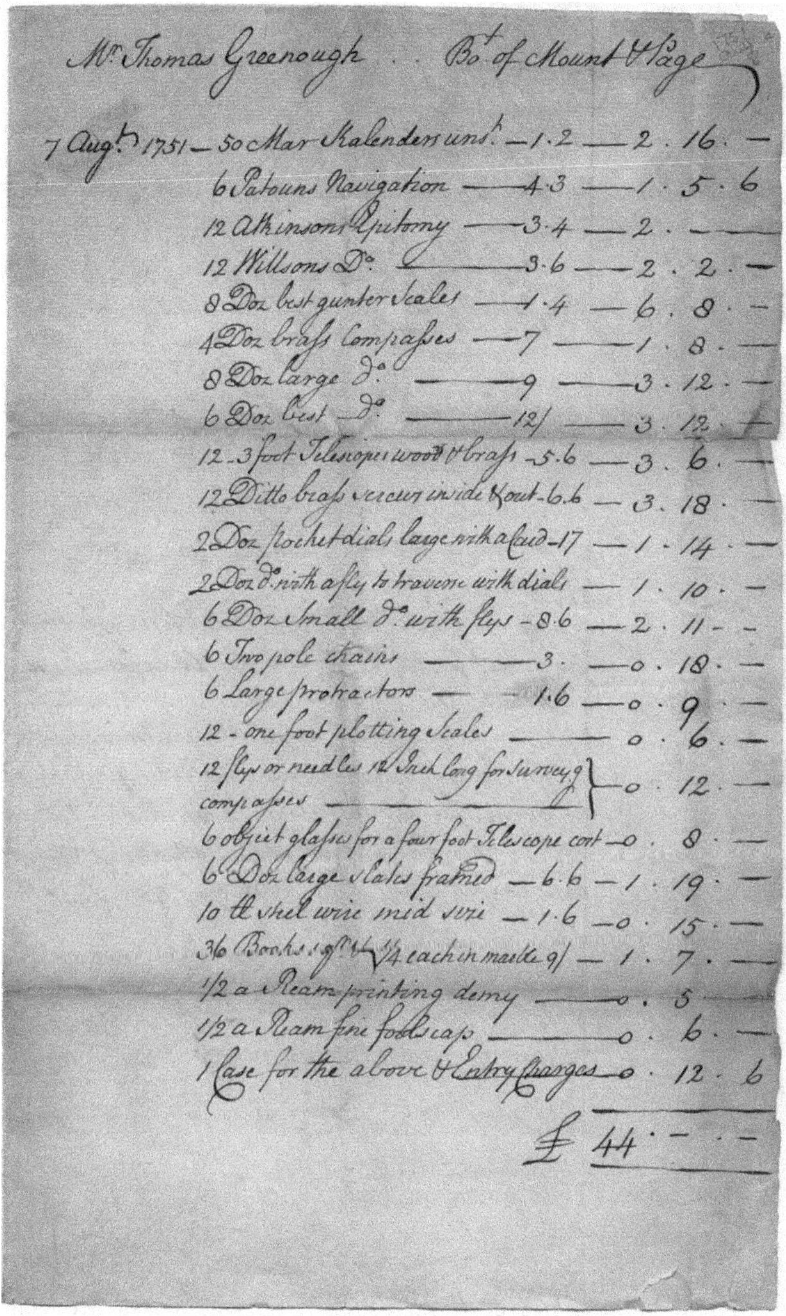
Mr Thomas Greenough . . Bot of Mount & Page

7 Augt 1751 — 50 Mar Kalenders unbd — 1.2 — 2 . 16 . —
6 Patouns Navigation — 4.3 — 1 . 5 . 6
12 Atkinsons Epitomy — 3.4 — 2 . — . —
12 Willsons Do — 3.6 — 2 . 2 . —
8 Doz best gunter Scales — 1.4 — 6 . 8 . —
4 Doz brass Compasses — 7 — 1 . 8 . —
8 Doz large Do — 9 — 3 . 12 . —
6 Doz best Do — 12/ — 3 . 12 . —
12 - 3 foot Telescopes wood & brass - 5.6 — 3 . 6 . —
12 Ditto brass screws inside & out - 6.6 — 3 . 18 . —
2 Doz pocket dials large with a Case - 17 — 1 . 14 . —
2 Doz Do with a fly to traverse with dials — 1 . 10 . —
6 Doz Small Do with flys - 8.6 — 2 . 11 . —
6 Two pole chains — 3. — 0 . 18 . —
6 Large protractors — 1.6 — 0 . 9 . —
12 - one foot plotting Scales — 0 . 6 . —
12 flys or needles 1½ Inch long for Surveyg compasses } — 0 . 12 . —
6 object glasses for a four foot Telescope cont — 0 . 8 . —
6 Doz large Slates framed — 6.6 — 1 . 19 . —
10 lb steel wire mid size — 1.6 — 0 . 15 . —
36 Books 1 qr & 1/4 each in marble 9/ — 1 . 7 . —
1/2 a Ream printing demy — 0 . 5 . —
1/2 a Ream fine foolscap — 0 . 6 . —
1 Case for the above & Entry Charges — 0 . 12 . 6
£ 44 . — . —

FIGURE 8.7 *Invoice, Mount & Page of London, for Thomas Greenough of Boston, 7 August 1751.*
DAVID S. GREENOUGH PAPERS, MASSACHUSETTS STATE HISTORICAL SOCIETY, MS. N-1355, BOX 5, FOLDER 1751.

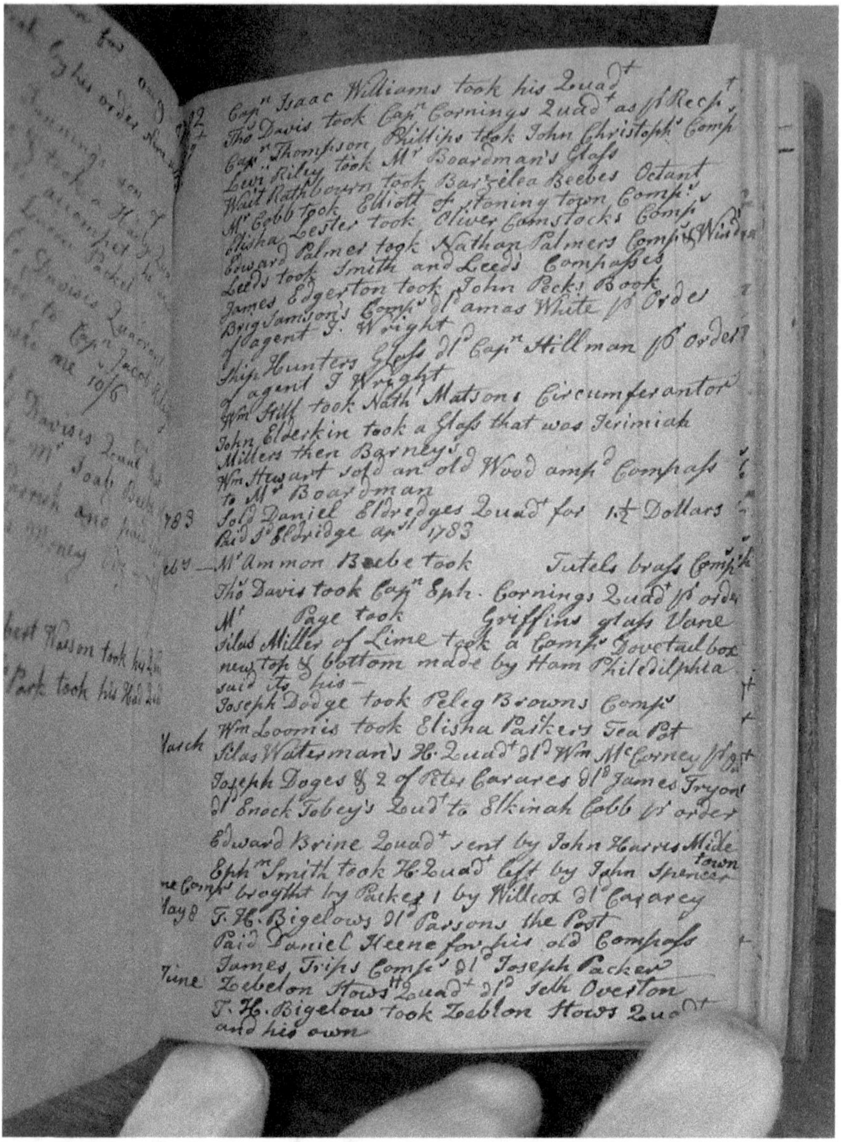

FIGURE 8.8 *Clark Elliott's account book for 1782–1783, f. 2r.*
NEW LONDON COUNTY HISTORICAL SOCIETY, MSBD 158.

made by Elliott, but does not record their sales. In the 1790 census records for Connecticut, I have been able to identify half of Elliott's recorded 377 buyers and sellers. Ninety-eight lived elsewhere in the colony (or state), mostly in towns along the Connecticut River (15 in Middleton, 15 in Weathersfield,

10 in Saybrook). I estimate that the average customer lived about 35–40 miles from New London (Connecticut was about 90 × 55 miles in size). If we assume that some (perhaps many?) of the unidentified names in Elliott's accounts lived outside Connecticut, we could conclude that he was serving a regional market extending well beyond the confines of New London.[87]

TABLE 8.4 *Clark Elliott's commercial transactions, 1771–1793. Key to column headers: Transactions, New, Consignment, Hadley's quadrant, Davis quadrant, Compass, Spyglass, Other*

Dates	Trans	New	Cons	HadQ	DavQ	Quad	Comp	SpyG	Other
1771–1774	9	1	8	3	2	3	0	0	1
1782–1785	97	14	83	16	4	23	40	3	11
1786–1789	63	15	48	17	5	21	12	5	3
1790–1793	44	21	23	21	6	6	6	3	2
Totals	213	51	162	57	17	53	58	11	17

Finally, I consider two clockmakers in smaller towns who also made and sold surveying instruments. Both men had been apprenticed to Thomas Harland (1735–1807) of Norwich, Connecticut, one of America's first watchmakers. Both men's account books reveal a trade based in barter rather than cash; both served as financial middlemen between other local craftsmen and retail customers; both offered all kinds of metal-working and finishing, mostly in brass and silver; and both repaired far more than they sold new. Despite their rural locations, both appear to have advertised; one placed an advertisement in the *Connecticut Courant* (Hartford) in 1791 and the other placed four advertisements in his local newspaper in the 1790s and issued a broadside in 1806.[88]

87 Elliott made his first 105 wooden quadrants during his apprenticeship with Greenough; he completed no. 253 in 1760, no. 348 in 1765, no. 398 in 1774, and no. 530, the final item recorded, in 1780, all in MsBd 158, op. cit. (n. 86), *passim*. For surviving Davis quadrants by Elliott, see Warner, "Davis' quadrants …", op. cit. (n. 8), pp. 29–30.

88 Daniel Burnap in *Connecticut Courant*, 4 April 1791; Jedidiah Baldwin in *Hampshire Gazette*, 6 July 1791; *Dartmouth Centenniel*, 28 October 1793 and 5 October 1795; *Dartmouth Gazette*, 22 August 1801; Trade broadside, 1806, Dartmouth College Library, Special Collections, Broadside 806224.

Daniel Burnap (1759–1838) of East Windsor, Connecticut (population 2,600 in 1790), sold about fifty high-end eight-day tall striking clocks over the last two decades of the eighteenth century. His accounts show that he also sold eleven surveyor's compasses, four to customers in East Windsor, six to customers more than twenty miles distant.[89] Jedidiah Baldwin (1768–1849) was the first clock-maker in Hanover, New Hampshire (population 1,400 in 1790), where he also served as Postmaster, Sealer of Weights and Measures, Sealer of Leather, Fence Viewer, Surveyor of Highways, and member of the School Committee. Before he left Hanover in 1811, Baldwin is known to have made fifty-five tall clocks, mostly purchased by people living within ten miles of Hanover. He also sold three surveyor's compasses and one brass compass card, each to customers living more than fifteen miles from Hanover.[90] The "proper scope" of a country instrument maker might have been one or two day's ride by horse.[91]

Appendix: Preliminary Database of Pre-1801 American 'Makers'

This database seeks to include all those who represented themselves as 'makers' of mathematical, philosophical or optical instruments in eighteenth-century America, either by signing an instrument extant today or by advertising or otherwise publishing notices to that effect. I have compiled names from several well-known studies of American instruments, indices of two relevant journals, *Rittenhouse* and the *Bulletin of the Scientific Instrument Society*, signed artifacts extant in American museums or private collections, databases of craftsmen or signed instruments, and an ever-growing

89 Penrose Hoopes, *Shop records of Daniel Burnap, clockmaker*, Hartford, 1958.

90 Jedidiah Baldwin Papers, Dartmouth College Library, Special Collections, 9 volumes of account books, Ms. 001152. For purchasers of Baldwin's clocks and instruments, see Frederick Chase, *A history of Dartmouth College and the town of Hanover, New Hampshire, edited by John K. Lord*, Cambridge, MA, 1891, *passim*; John K. Lord, *History of the town of Hanover, N.H.*, Hanover, 1928, pp. 33–34; Philip Zea, "Clockmaking and society at the River and the Bay: Jedidiah and Jabez Baldwin, 1790–1820", in Peter Benes and Jane M. Benes (eds.), *The Bay and the River, 1600–1900: Dublin Seminar for New England Folklife Annual Proceedings, 1981*, Boston, 1982, pp. 46–59; Charles S. Parsons, "Jedidiah Baldwin: Clockmaker & silversmith when at Hanover, New Hampshire, and his nine volumes of account books, 1794–1810", 1987, typescript, Dartmouth College Rauner Special Collections Library, D.C. Hist TS544.8 .B35 1987.

91 According to Shammas, op. cit. (n. 5), p. 272, colonial New England farmers "travelled on average 25 miles or more to sell their crop or livestock", which suggests that "they were accustomed to dealing with more than just the local merchandiser and relied on different people for selling and for buying".

database of scanned newspapers published in North America since 1690.[92] I have consulted some (but not all) published city directories (which start only in the 1780s), but no local tax, court, or census records. Inspired by Gloria Clifton's *Directory of British scientific instrument makers, 1550–1851* of 1995, I have sought to follow her selection criteria but have not culled the American sources as thoroughly as she did the British sources.[93] This database, thus, must be regarded as preliminary.

The list presents 180 names, birth and death dates when known, dates and locations where they are thought to have begun their 'making' activities, and the sources (followed by page numbers) in which I have found this information. It does not record shifts in location over a career. Occasionally I cite additional sources on given makers, but I have not systematically attempted to locate all published studies of these individuals. My goal in assembling this database is not completeness but rather to construct a representative sample in which patterns can be explored.

Maker	Date	Location	Sources
Avery, John (1762–1834)	1800	Bridgewater, NY	S183
Badger, Joseph	1784	New Haven, CT	W
Bailey, John (1731–1823)	1760	Hanover, MA	S4
Bailey, John II (1752–1823)	1770	Hanover, MA	B39
Baily, Joel (1732–1797)	1764	West Bradford, PA	W
Balch, Thomas H.	1771	Newburyport, MA	W, AHN
Baldwin, Jedidiah (1768–1849)	1794	Hanover, NH	B123, AHN

92 Basic sources include AHN = *America's Historical Newspapers*, op. cit. (n. 30); ANB = *American National Biography*, New York, 2000; B = Bedini, *Early American scientific instruments…*, op. cit. (n. 9); Br = Brewington, op. cit. (n. 8); D(year) = Published city directory for the year listed; G = James L. Garvin and Donna-Belle Garvin, *Instruments of change: New Hampshire hand tools and their makers*, Canaan, NH, 1984; Gi = Harrold E. Gillingham, "Some early Philadelphia instrument makers", *Pennsylvania magazine of history and biography* 51 (1927), pp. 289–307; H = Harris, op. cit. (n. 30); L = JEFF LOCKE, "Surveying America: The art of the instrument", *Antiques journal* (April 2006), pp. 36–40; NMAH = National Museum of American History, Surveying and Geodesy, http://amhistory.si.edu/surveying/; S = Charles E. Smart, *The makers of surveying instruments in America since 1700*, 2 vols., Troy, NY, 1962–1967; W = Webster Signature Database, http://historydb.adlerplanetarium.org/signatures/; Wq = WARNER, "Davis' quadrants…", op. cit. (n. 8); Ws = D. J. Warner, "American octants and sextants: The early years", *Rittenhouse* 3 (1989), pp. 86–112. I have also consulted the Museum of Early Southern Decorative Arts Craftsman Database, http://research.oldsalemonline.org/ProficioPublicSearch/CollectionsHome.aspx?CRAFTSMAN (all websites accessed 15 March 2016).

93 See Clifton, op. cit. (n. 3), pp. xii–xv. I exclude self-designated watch- and clockmakers, globe-, map- and almanac-makers, spectacle makers, and makers of medical or surgical apparatus from this database.

Ballard, Samuel (1718–1793)	1750	Boston, MA	S5
Bennett, Nehemiah	1777	Middleboro, MA	B131, W
Biddle, Owen (1737–1799)	1764	Philadelphia, PA	B21–22
Biggs, Thomas	1786	New York, NY	B59, S9, Ws90, AHN
Blakslee, Ziba (1768–1834)	1792	Newton, CT	B47
Blundy, Charles	1753	Charleston, SC	B29
Bond, William	1793	Boston, MA	ANB
Bowditch, Nathanial (1773–1838)	1792	Boston, MA	Br121, W
Buell, Abel (1742–1822)	1800	New Haven, CT	S189
Burges, Bartholomew	1789	Boston, MA	B40
Burnap, Daniel (1759–1838)	1780	Windsor, CT	B69, AHN[94]
Callender, Joseph	1780	Boston, MA	NMAH
Caritat, H.	1799	New York, NY	B51
Chandlee, Benjamin, Jr. (1723–1791)	1780	Nottingham, MD	B54
Chandlee, Ellis (1755–1816)	1790	Nottingham, MA	B55, S26, AHN
Chandlee, Goldsmith (1746–1821)	1783	Winchester, VA	NMAH, S26
Chandlee, Isaac (1769–1813)	1780	Nottingham, MD	S29, AHN
Chandler, Timothy (1762–1848)	1792	Concord, NH	G69, S190
Chase, James (1737–1812)	1760	Gilmanton, NH	W
Chickering, Nathaniel (1750–1837)	1780	Dover, MA	S29
Cilfton, George	1800	New York, NY	W
Churchman, John (1753–1805)	1785	Notthingham, MD	W
Claggett, William (1696–1749)	1724	Newport, RI	AHN[95]
Clark, Horatio (1772–1883)	1796	Bennington, VT	S30
Clark, Robert	1785	Charleston, SC	B31, AHN
Clough, Jere	1760	Boston, MA	B105
Clough, Samuel (1665–1707)	1707	Boston, MA	A1707
Colles, Christopher	1774	New York, NY	W
Condy, Benjamin (d. 1798)	1763	Philadelphia, PA	Ws93, D1785, Gi293
Coram, Robert (d. 1796)	1790	Wilmington, DE	H77, AHN
Crow, George (1725–1772)	1754	Wilmington, DE	B54
Cutler, J.	1777	Salem, MA	W
Dabny (Dabney), John	1739	Boston, MA	B27, AHN
Dakin, Jonathan	1745	Boston, MA	B38

94 Hoopes, op. cit. (n. 89).

95 Champlin, op. cit. (n. 69), pp. 157–185.

Danner, Jacob (1765–1850)	1790	Middleton, VA	S35
Davenport, Michael	1791	Philadelphia, PA	B61
Dean, William (d. 1797)	1792	Philadelphia, PA	B60, S195, Ws94, AHN
Denegan, John	1785	Philadelphia, PA	Gi305, W, AHN
Devacht, François & Joseph	1792	Gallipolis, OH	B48, W
Dickinson, Edmund	1772	Williamsburg, VA	W, AHN
Dobson	1798	Philadelphia, PA	W
Dod, Daniel (1778–1823)	1800	Mendham, NJ	S195
Dod, Lebbeus (1739–1816)	1760	Mendham, NJ	NMAH, W
Donegany, Joseph	1787	New York, NY	B33
Doolittle, Enos (1751–1806)	1788	Hartford, CT	S37, AHN
Doolittle, Isaac I (1721–1800)	1763	New Haven, CT	B45
Doolittle, Isaac II (1759–1821)	1780	New Haven, CT	B45, S38
Dorsey, Leven	1795	Norfolk, VA	H20, AHN
Dorsey, Philip	1796	Baltimore, MD	S39, Ws95
Dring, Thomas	1796	West Chester, PA	B31
Drury, Stephen (c. 1767-c. 1851)	1788	Huntingdon Co., PA	W
Dudley, Benjamin	1768	Newport, RI	H39, AHN
Dunbar, John	1796	Baltimore, MD (?)	H120, AHN
Dupee, John (b. 1729)	1756	Boston, MA	B104, S40, Br125, L26
Ellicott, Andrew (1754–1820)	1774	Philadelphia, PA	B19, NMAH
Elliott, Clark (1732–1793)	1767	New London, CT	Wq29, W, AHN
Evans, George (d. 1798)	1796	Philadelphia, PA	Gi304, B31, AHN
Ewing, John (1732–1802)	1750	Philadelphia, PA	W, Ws96
Fenton, Jonathan	1790	New Haven, CT	AHN[96]
Fisher, John	1790	Yorktown, PA	W
Fisher, Martin	1791	Philadelphia, PA	Gi306, W, AHN
Folger, Walter Jr. (1765–1849)	1790	Nantucket, MA	B40
Ford, George (d. 1842)	1800	Lancaster, PA	B63
Franklin, Benjamin (1709–1790)	1740	Philadelphia, PA	W
French, James	1788	Providence, RI	Wq31
Gatty, Henry	1794	New York, NY	Ws96, AHN

96 Schechner, "Telescopes in colonial and federal America ..." op. cit. (n. 83).

Gatty, Joseph	1794	New York, NY	Gi306, B33, AHN
Gilman, Benjamin C. (1763–1835)	1790	Exeter, NH	B34, S200[97]
Gilmur, Bryan	1790	Philadelphia, PA	B63, W
Godfrey, Thomas (1704–1749)	1730	Philadelphia, PA	Gi291, B58, Ws97
Goff, William	1782	Georgetown, SC	W
Gould, Chester	1794	Philadelphia, PA	Br127
Gould, John (d. 1797)	1794	Philadelphia, PA	B30, Ws98, AHN[98]
Greenleaf, Stephen (d. 1752)	1745	Boston, MA	B37, AHN
Greenough, Thomas (1710–1785)	1742	Boston, MA	B85, Wq31, AHN
Greenough, William	1785	Boston, MA	W
Greenwood, Clark (d. 1810)	1800	New York, NY	D1800, AHN
Greenwood, Isaac Jr. (1730–1803)	1760	Cambridge, MA	B38, Wq32, Ws99, AHN
Greenwood, John	1786	New York, NY	W, AHN
Griffith, Nathaniel	1770	Portsmouth, NH	W
Hagger, Benj. King (1769–1834)	1787	Baltimore, MD	B109, Br128, Ws100
Haggar, Jonas	1799	New London, CT	H, AHN
Haggar, William G. (1744–1830)	1760	Newport, RI	B44, Wq32, AHN
Halsy, James I (d. 1696)	1767	Boston, MA	Wq34, Br128[99]
Halsy, James II (1695–1766)	1739	Boston, MA	B84, Br29, AHN
Halsy, John	1702	Boston, MA	B82
Halsy, Joseph I (1657–1745)	1738	Boston, MA	B80, Wq34, Br129
Halsy, Joseph II (b. 1699)	1738	Boston, MA	Br129, AHN
Ham, Anthony	1780	Philadelphia, MA	W, L26
Ham, James I	1754	New York, NY	B65, Wq35, Ws100, AHN

97 Frank O. Spinney, "An ingenious Yankee craftsman", *The magazine antiques* 44 (1943), pp. 116–119.

98 Clifton, op. cit. (n. 3), p. 116.

99 Bedini, "At the sign of the compass and quadrant…", op. cit. (n. 11), p. 27.

Ham, James II	1780	Philadelphia, PA	B66, D1785, Wq35, AHN
Hamlin, William (1772–1869)	1800	Providence, RI	B44
Hanks, Benjamin (1755–1824)	1777	Litchfield, CT	S70
Harland, Thomas (1735–1807)	1790	Norwich, CT	NMAH
Hart, William (1734–1812)	1753	Portsmouth, NH	S79, G76, Wq35
Hassam, Stephen (1761–1861)	1788	Charlestown, NH	S79
Hellig, John (1765–1841)	1790	Germantown, PA	S207
Heisely, Frederick (1759–1839)	1780	Fredericktown, MD	S81, NMAH, AHN[100]
Hempstead, G.	1780	Walpole, MA	W
Hicks, Adrian B. K.	1797	New York, NY	W, D1800
Hinton, William (d. 1784)	1772	New York, NY	B66, Wq36, Ws101, AHN[101]
Hobbes, James	1796	Baltimore, MD	S84
Hoff, George (1740–1816)	1770	Lancaster, PA	B157, S84, AHN
Hoff, John (1776–1819)	1800	Lancaster, PA	S214
Houghton, Rowland (1678–1744)	1735	Boston, MA	S88, AHN[102]
Huntington, Gurdon (1763–1804)	1789	Walpole, NH	B118, G78, S89
Hurd, Nathaniel	1780	Boston, MA	W
Hurtin, Christian (1743–1830)	1792	Goshen, NY	S216
Hutchins, Levi (1761–1855)	1786	Concord, NH	G78, S219
Ingham, North	1730	Boston, MA	S90
Jacks, James	1797	Philadelphia, PA	B63
Jay, Daniel	1764	Philadelphia, PA	W
Keappock, James	1753	Philadelphia, PA	W, Ws102
Ketterer, Alloysius	1789	Philadelphia, PA	Gi305, W, AHN
King, Benjamin I (1707–1786)	1759	Newport, RI	B132, Wq36, Ws102

100 Bedini, "Frederick Heisely ...", op. cit. (n. 80), pp. 120–124.

101 Bedini, "At the sign of the compass and quadrant ...", op. cit. (n. 11), p. 62.

102 S. A. Bedini, "Rowland Houghton's 'new theodolate'", *Rittenhouse* 1 (1987), pp. 30–39; Raymond V. Giordano, "Some notes on the two extant Rowland Houghton new theodolates", *Rittenhouse* 15 (2001), pp. 93–97.

King, Benjamin II (1740–1804)	1765	Salem, MA	B132, B37, Ws103
King, Daniel (1704–1790)	1744	Salem, MA	B133, B36, S93, AHN
King, Gedney (1777–1839)	1800	Boston, MA	B133, S94, Ws104, AHN
King, Samuel (1748–1819)	1786	Newport, RI	S95, AHN
Lamb, Anthony (1703–1784)	1730	New York, NY	B133, Ws105, AHN[103]
Lamb, John (1735–1800)	1760	New York, NY	B29
Long, Enoch	1763	Hopkinton, NH	W
Main, David Jr.	1800	Connecticut?	L
Marshall, John (1730–1802)	1772	Fauquier, VA	W, B8
Mcallister, William Y.	1783	Philadelphia, PA	W
Mendenhall, Thomas	1775	Lancaster, PA	S228, AHN
Michael, Lewis	1780	?	L
Miller, Aaron (d. 1771)	1748	Elizabethtown, NJ	B53
Newell, Andrew (1749–1798)	1789	Boston, MA	B106, S112, AHN
Newell, Joseph	1800	Boston, MA	B107, S113
Noble, Philander (1772–1845)	1800	Westfield, MA	S232
Platt, Benjamin (1757–1833)	1780	Litchfield, CT	B49, S124
Pope, Joseph (1750–1826)	1787	Boston, MA	B39
Potter, John (1746–1818)	1785	Brookfield, MA	B43, S128[104]
Potts, William Lukens	1800	Bucks, PA	S129, NMAH
Prince, John (1751–1836)	1780	Salem, MA	B25
Pryor, Thomas	1760	Philadelphia, PA	B59, D1785, Gi303, AHN
Quincy, J.	1780	New York, NY	Ws107
Revere, Paul (1735–1818)	1786	Boston, MA	W
Richardson, William	1796	Philadelphia, PA	D1796, AHN
Rittenhouse, Benj. (1740–1820)	1780	Philadelphia, PA	S136
Rittenhouse, David (1732–1796)	1770	Philadelphia, PA	S138, D1785, AHN[105]

103 Bedini, "At the sign of the compass and quadrant…", op. cit. (n. 11), p. 2 ff.

104 William H. Skerritt (ed.), *Catalog of the Charles E. Smart collection of antique surveying instruments*, Troy, 1996, pp. 98–100.

105 Brooke Hindle, *David Rittenhouse*, Princeton, 1964.

Rivington, James	1778	New York, NY	W, AHN
Sadtler, Philip B.	1800	Baltimore, MD	W
Scott, John (1752–1801)	1780	Chambersburg, PA	S243
Smith, Cordial	1775	Connecticut?	S156, W
Sommer, Mrs. Balthasar (widow)	1751	New York, NY	W, AHN[106]
Southworth, Elijah	1797	New York, NY	W, D1800
Sower, Christopher (1693–1740)	1730	Germantown, PA	B63
Stewart, James	1792	Baltimore, MD	AHN
Stobbart, Thomas	1735	Boston, MA	AHN
Storrs, Nathan (1768–1839)	1795	Northampton, MA	S249, L
Sweeney	1800	Baltimore, MD	AHN
Taws, Charles	1795	Philadelphia, PA	Gi303, B61, D1795
Thaxter, Samuel (1769–1842)	1792	Boston, MA	S160, Br141, Ws111, AHN
Trundy, John (1774–1802)	1800	Portsmouth, NH	S160
Walker, Ralph	1793	Jamaica	Br143
Wall, George Jr.	1788	Bucks, PA	B63
Walpole, Charles	1740	New York, NY	B28, AHN
Warren, Benjamin (1740–1825)	1786	Plymouth, MA	B112, S255
Welker, John	1797	Washington, DC	W
Wheaton, Seth	1799	Providence, RI	W
Whidden, Michael (1695–1773)	1757	Portsmouth, NH	G94
White, George	1800	Walpole, NH	G94
White, Peregrine (1747–1834)	1800	Woodstock, CT	B47, NMAH
Whitney, Thomas (d. 1823)	1798	Philadelphia, PA	Gi304, S166, NMAH, AHN
Willard, Alexander (1774–1850)	1800	Ashby, MA	S260
Williams, Benjamin	1798	Portland, ME	AHN
Williams, David	1780	Newport, RI	W
Williams, William (1738–1792)	1768	Boston	B93, Br144, AHN
Wilstar, Richard	1752	Wistarburg, NJ	B53
Witt, Christopher (1675–1765)	1710	Germantown, PA	B62[107]
Young, Frederic	1792	Albany, NY	AHN

106 Maria Rooseboom, *Bijdrage tot de Geschiedenis der Instrumentmakerskunst in de noordelijke Nederlanden tot omsteeks 1840*, Leiden, 1950, p. 125; Zuidervaart, op. cit. (n. 82), pp. 95–96; and Schechner, "Telescopes in colonial and federal America …", op. cit. (n. 83).

107 Hocker, op. cit. (n. 9).

CHAPTER 9

Trade in Medical Instruments and Colonialist Policies between Mexico and Europe in the Nineteenth Century

*Laura Cházaro**

Introduction

The overwhelming materiality of medical and scientific instruments tends to impose itself as the 'natural' destiny of modern medical knowledge and practice, to such a degree that many historical analyses examine those objects as if they were independent of their places of use and variations and peculiarities of context; that is, as if they were subordinate to theories or 'reified theorems' (*reifiés*) as defined by Gastón Bachelard.[1] Such approaches to the travels of medical instruments tend to view them as simple objects involved in a 'natural' expansion of proven theories—usually from Europe to its colonies—justified by a supposed endemic absence of the engineering knowledge required to produce them *in situ*. Their uneven circulation in markets is explained mainly by this ahistorical materiality and diffusionist theses which presuppose that scientific objects and the knowledge associated with them spread from an enlightened, industrialized Europe to a pre-capitalist America virtually devoid of science.[2] However, historical evidence exists that the intense exchange of

* I wish to thank the editors for their insightful suggestions to an earlier version of this paper, which provided an opportunity to rethink some of my arguments. Chapter translated by Paul Kersey.

1 Bachelard states: "In modern science, an instrument is actually a reified theorem; if we take the schematic construction of experience piece by piece, or even instrument by instrument, we can see that the hypotheses must be coordinated from the point of view of the instrument itself; apparatuses such as Millikan's [...] are conceived directly as a function of the electron and the atom"; Gaston Bachelard, *Epistemología*, Barcelona, 1973, p. 164. For a critical review, see Simon Schaffer, "Natural Philosophy", in George S. Rousseau and Roy Porter (eds.), *The Ferment of Knowledge: Studies in the Historiography of Eighteenth-century Science*, Cambridge, 1980, pp. 77–91.

2 For a classical example of this view, see George Basalla, "The Spread of Western Science: A three stage model describes the introduction of modern science into any non-European nation", *Science* 156 (May 1967), pp. 611–622.

 | DOI 10.1163/9789004324930_010

scientific devices that occurred across the Atlantic also mirrored an exchange of scientific knowledge and practice.[3] Today the literature contains several important case studies that debate the characteristics and forms of these scientific exchanges in the Americas.[4]

The question is how to demonstrate this history without glossing over the reality that Mexico *did not* have a vigorous industry in scientific instruments, nor even the basic production machinery required to manufacture them? I began this inquiry by tracing the commercial exchanges between Mexico and Europe, which brought ready-made apparatus for cardiographic readings to the Americas, including the sphygmograph (*Fig. 9.1*) designed by the French physiologist Étienne Jules Marey (1830–1940).[5] Within a thirty year period—between 1833 and 1863—such instruments would flood classrooms of clinical medicine at the National School of Medicine (NSM) in Mexico City and then the laboratories of the National Institute of Medicine (NIM), founded in 1889, which played a prominent role in the research of Dr. Daniel Vergara Lope (1865–1938) into the normalcy of high-altitude respiration among the inhabitants of Mexico's high central valley, situated at 7,349 feet above sea level.

3 This study was inspired by several readings: Simon Schaffer and Steven Shapin, *The Leviathan and the Air Pump: Hobbes, Boyle and the Experimental Life*, Princeton, 1985; Marie Noelle Bourguet, Christian Licoppe, and H. Otto Sibum (eds.), *Instruments, Travel and Science: Itineraries of Precision from the Seventeenth to the Twentieth Century*, London, 2003; Albert Van Helden and Thomas L. Hankins, "Introduction: Instruments in the History of Science", *Osiris* 9 (1994), pp. 1–6: 3; Antonio Lafuente, "Enlightenment in an Imperial Context: Local Science in the Eighteenth Century", in Roy MacLeod (ed.), "Nature and Empire: Science and the Colonial Enterprise", *Osiris* 15 (2000), pp. 155–73; and Willem D. Hackmann, "Scientific Instruments: Models of Brass and Aids to Discovery", in David Gooding, Trevor Pinch and Simon Schaffer (eds.), *The Uses of Experiment. Studies in Natural Sciences*, New York, 1999, pp. 31–65.

4 For an alternative to the diffusionist view see: James Blaut, *The Coloniser's Model: Geographical Diffusionism and Eurocentric History*, New York, 1993; Roberto Schwartz, "Ideas fuera de lugar", *Meridional. Revista Chilena de Estudios Latinoamericanos* 3 (2014), pp. 183–199; Ana Carolonia Vimeiero Gomes, "'Too Good to be True': The Controversy over the Use of Permanganate of Potash as an Antidote to Snake Poison and the Circulation of Brazilian Physiology in the Nineteenth Century, *Bulletin of the History of medicine* 86 (2012), pp. 153–177.

5 Predecessors of Marey's sphygmograph were the myograph by Hermann Von Helmholtz (1821–1894), a device designed to detect electrical discharges produced by muscular contractions, and Vierordt's famous "pulse writer" (1855). The use of these devices soon spread to other areas of physiological research and eventually led to Marey's sphygmograph, introduced in 1859 in *Recherches sur la circulation de sang à l'état sain et dans les maladies*, and later, in a perfected version, in Jules E. Marey, *La Méthode Graphique dans les Sciences Expérimentales*, G. Masson Éditeur, Paris, 1878.

FIGURE 9.1 *Representation of Marey's sphygmograph in a Mexican book specialized in cardiac ailments.*
EDUARDO HERNÁNDEZ, DE LA UTILIDAD DEL ESFIGMÓGRAFO PARA EL DIAGNÓSTICO DE LAS LESIONES ORGÁNICAS DEL CORAZÓN, *MÉXICO, 1873, P. 18.*

Decisions concerning the instruments to be used, the means of obtaining them, and the types that should be reproduced, all depended on both scientific and political interests. The circulation of scientific devices during the nineteenth century seems to have been governed by a postcolonial dynamics, which we can describe as 'export-import growth'; that is, exports of raw materials and imports of some machinery, including scientific instruments. My concern here is not to explain the global relations between exports and imports—or production—but to describe how the Mexican medical elite consumed scientific instruments as foreign commodities that produced both knowledge and politics:[6] instruments that generated stereotypes while also influencing the emergence of nationalist policies, as a local medical elite recreated European instruments but adapted them to their own ideas and findings. Hence, the absence of domestic manufacturing of medical instruments was a void filled by the policies of state officials and physicians whose importation and artisanal re-creation of such devices allowed them to be used, *ad hoc*, as

6 Benjamin Orlove and Arnold J. Bauer, "Giving Importance to Imports", in B. Orlove (ed.), *The Allure of the Foreing: Imported Goods in Postcolonial Latin America*, Ann Arbor, 1997, pp. 1–29.

objects and knowledge to further their own political interests. In this case, the act of reproducing instruments—for artisanal not commercial purposes—became important not only as a means of acquiring knowledge but also as a political gesture. Neither poverty nor a lack of ingenuity suffices to explain why those involved in the medical-instrument trade between Mexico and Europe made no effort to convert them into commodities; i.e., objects of national production. Instead, they incorporated them as artisanal products because their peculiar, artistic design allowed them to produce, simultaneously, scientifically valid knowledge and a set of stereotypes of national identification, such as the 'normality and modernity of the Mestizo'. I intend to show that commercial exchange between Europe and Mexico affected instruments and subjects mutually, as neither one, once mobilized, could remain intact or identical, but became attuned or calibrated according to nationalist and local styles.[7]

Sphygmographs: From France to Mexico

Few scientific instruments circulated in early nineteenth century Mexico, but most of those that did were bought in Europe and destined for private practice. This began to change with the creation in 1838 of the National School of Medicine, which required imported instruments for teaching and hospital use. The National School of Medicine's archives contain records of large orders of artefacts for classes on Clinical Medicine, Surgery and Obstetrics, including stethoscopes, laryngoscopes, forceps, surgical instruments, and anatomical specimens.[8] For courses on Physiology and Therapeutics, imported materials

7 I have examined other aspects of Vergara Lope's work in: Laura Cházaro, "Medición y estadísticas: La fisio-antropometría de la respiración en las alturas, un debate por la patria", *Ciencias. Revista de Difusión de la Facultad de Ciencias de la UNAM* 60–61 (2001), pp. 37–43, and L. Cházaro, "Midiendo las diferencias, buscando la igualdad. Un ensayo sobre la fisiología en México a fines del siglo XIX", in Javier Pérez Siller and Chantal Cramaussel (eds.), *México-Francia. Memoria de una sensibilidad, Siglos XIX–XX, Tomo II*, México, 2004, pp. 317–339.

8 I make no attempt—as it would be impossible—to give a complete list of the enormous variety of instruments, artifacts, and utensils that circulated in the ENM, NIM and Hospitals. I have examined inventories from the following archives and repositories: Archivo Histórico de la Facultad de Medicina-Anexo Fondo Escuela de Medicina y Alumnos (hereinafter: AHFM-AFEMyA); Archivo General de la Nación, Instrucción Pública y Bellas Artes, Instituto Médico Nacional (AGN-IPyBA, IMN) and the Secretaría de Salubridad y Asistencia, Fondo Beneficencia Pública, Hospitales (AHSS-BP-Hospitales). In 1882, the budget requested by Dr. Capetillo for the Course on Clinical Obstetrics and Obstetric Theory made up 53% of the total amount solicited for twelve courses (12,508 French francs); while the amount requested for Physiology represented only 26% (471 French francs).

included equipment for chemical analyses and, above all, instruments that produced graphical readings: cardiographs to record heart movements (systole, diastole), sphygmographs to chart changes in blood pressure and pulse when placed on a person's wrist, myographs to measure muscle contractibility and elasticity, and pneumographs to indicate amounts of respired air.[9]

In terms of quantity, graphical recording devices were the most numerous instruments imported to Mexico. Though records name several makers, most models were developed by the aforementioned Marey, who was especially interested in the study of movement and designed devices to analyse muscle action and cardio-respiratory and nervous functions.[10] The sphygmograph—designed to record heart movements without vivisection or mutilation—had a mechanism with a needle that enabled it to trace a subject's heartbeat on smoked paper. Marey held that his graphs were precise records, superior to the subjective judgments of clinical observers because they translated movement into curves, which could be reproduced and measured at will.[11] Those lines seemed capable of eliminating all bias or error, as they apparently required neither doctors nor interpretative acumen.[12]

Francisco Ortega, "Presupuesto de instrumentos que se necesitan para las cátedras", 7 February 1882: México, AHFM-AFEMyA, file 674, fols. 1–3.

9 Examples of the purchase and circulation of sphygmographs and other devices related to observing cardio-respiratory phenomena include: "Carta del Director de la ENM al Ministro de Relaciones", 1849: AHFM-AFEMyA, file 270; "Inventarios", 1872–1873: AHFM-AFEMyA, files 534, 26 fols. and file 497, 24 fols. For cases of instrument purchases at the NIM, see also note 14.

10 On Marey see Martha Braun, *Picturing Time: The Work of Etienne-Jules Marey (1830–1904)*, Chicago, 1992; François Dagognet, *Etienne Jules Marey: A Passion for the Trace*, New York, 1992, p. 30 and ff.; on physiology, Marey and experimentation, see Timothy Lenoir, "Laboratories, medicine and public life in Germany, 1830–1849: ideological roots of the institutional revolution", in Andrew Cunningham and Perry Williams (eds.), *The Laboratory Revolution in Medicine*, Cambridge, 1992, pp. 14–71; Robert G. Frank Jr., "The Telltale Heart: Physiological Instruments, Graphic Methods and Clinical Hopes, 1854–1914", in William Coleman and Frederic L. Holmes (eds.), *Experimental Physiology in Nineteenth-century Medicine*, Berkeley, 1988, pp. 211–290; Soraya De Charadevian, "Graphical Method and Discipline: Self-recording Instruments in Nineteenth-century Physiology", *Studies in History and Philosophy of Science* 24 (1993), pp. 267–291; Stanley J. Reiser, *Medicine and the Reign of Technology*, Cambridge, 1978; and Merriley Borell, "Instrumentation and the Rise of Modern Physiology", *Science and Technology Studies* 5 (1987), pp. 53–62: 54.

11 Marey, op. cit. (n. 5), p. 1.

12 Lorraine Daston, "Objectivity and the Escape from Perspective", *Social Studies of Science* 22 (1992), pp. 597–618, has analyzed this distance as the "escape from perspective", the genesis of the notion of mechanical objectivity.

The same convictions concerning objectivity accompany the sphygmographs that came to Mexico, but once in the hands of Mexican doctors, these and similar recording instruments were used in local contexts. Though imported as teaching aids for clinical medicine, they came to enjoy special success in diagnosing respiratory and cardiac ailments. In the hands of some American and British doctors, sphygmographs were channelled into local clinical work, quite distinct from the tasks of physiological experimentation for which their French creators had designed them. In this regard we can refer to the experiences of Dr. Rafael Lavista with the sphygmograph, performed at the Hospital de San Andrés around the 1870s.[13]

It was not until the late nineteenth century that such instruments were incorporated into experimental work, when they were ordered by laboratories at institutions such as the National Institute of Medicine, which housed Mexico's first experimental physiology laboratory. Its inventories show that over 120 artefacts were acquired in less than five years, most of them graph-producing types.[14] Once in use, those devices showed themselves to be of a materiality that had no fixed uses, in the sense that their physical form did not fully determine the practices in which they were utilized. In Mexico, they moved from clinical spaces to physiology laboratories according to changes in their users' practices.

---

13 About the English case, see among others: Christopher Lawrence, "Alfred Henry Garrod and the Indirect Measurement of the Isometric Period of the Heart's contraction", *Medical History* 24 (1980), pp. 342–346. For the clinic use of the sphygmograph in Mexico, see: Rafael Lavitsa, *Importancia del método gráfico para el estudio de las funciones de la vida, Thesis*, México, 1870, pp. 18–19 and Eduardo Hernández, *De la utilidad del esfigmógrafo para el diagnóstico de las lesiones orgánicas del corazón*, México, 1873, p. 6.

14 In 1890, the NIM was organized in five sections: the Physiology Section and the Experimental Physiology Laboratory, Natural Medical History, Analytical Chemistry, Clinical Therapeutics, and Climatology and Geographical Medicine. To reconstruct the travels of apparatus like sphygmographs, cardiographs and polygraphs in general at the NIM, see the acquisitions inventories and correspondence on purchases and sales; eg., "Inventario Número 1 de los instrumentos, aparatos y útiles existentes en la Sección 3ª del Instituto Médico Nacional y que proceden del primer pedido que se hizo a Europa", 1 April 1891: AGN, IPyBA, IMN, box 124, file 4, fols. 62–176; and "Inventario de los útiles, muebles, enseres, aparatos, sustancias químicas, 3ª Sección del Instituto Médico Nacional", July 1898: AGN, IPyBA, IMN, box 124, file 4, fols. 253–272. On Vergara Lope and the NIM, see: Marcial Avendaño, "Antecedentes del IMN y los Primeros Años de Trabajo de la Sección 3a de Fisiología", *Boletín de Historia y Filosofía de la Medicina* 10 (2007), pp. 21–27; Ana C. Rodríguez de Romo, "Daniel Vergara Lope and Carlos Monge Medrano: Two Pioneers of High Altitude Medicine", *High Altitude Medicine and Biology* 3 (2001), pp. 299–309.

The Sphygmograph in the Tropical Mountains

The National Institute of Medicine, founded in 1889, was entrusted with the task of compiling a national pharmacopeia and determining the curative powers of medicinal plants, though in fact it supported other kinds of research, including Vergara Lope's study of high-altitude respiration. While not new, this topic became imbued with medical and political significance in 1861 when the French physician Dénis Jourdanet (1815–1892) published his theory of 'anoxihemia'—high-elevation anaemia—arguing that because the oxygen in the air of Mexico's high central plateau (above 7,400 feet) was rarefied, it predisposed Mexicans to anoxihemia and condemned them to live as "a miserable race".[15] When Jourdanet's ideas were seconded by the renowned French physiologist Paul Bert (1833–1886), a disciple of Claude Bernard, many Mexican doctors took them as authoritative. In order to study the effects of altitude on human health, Dénis Jourdanet and Paul Bert—like Vergara sometime later—utilized sphygmographs, cardiographs, and air pumps. Unlike Vergara, however, those French scientists never went to the mountains; rather, they performed their experiments at the Laboratory of the *Sorbonne*, right at sea level. To reproduce rarefied atmospheres they used a double human hypobaric chamber with a low pressure chamber. It is important to note that Jourdanet did recognise the effects of the malady known today as mountain sickness (*mal de montaña*), as he lived in Mexico for fifteen years, from 1842 to 1846 and from 1848 to 1860.

In his efforts to refute the anoxihemia thesis, Vergara used a multiplicity of instruments: cardiographs, sphygmographs, pneumographs, oxygenographs, and even a chamber for "compressed air baths" (oxygen rarefaction) called the Legay apparatus after its French creator.[16] Unlike many contemporary clinical physicians, Vergara was convinced that "the exactness of results" could not be entrusted "to the senses" but only to instruments used in the proper context, for example, on the slopes of the Popocatepetl or Iztaccihuatl volcanoes.[17]

15 Daniel Vergara Lope, *Refutación teórica y experimental de la teoría de la Anoxihemia del Doctor Jourdanet* (México, Inaugural Thesis), México, 1890, pp. 12 and 34. See also: Denis Jourdanet, *Les altitudes de l'Amérique tropicale Comparée aux Niveaux des Mers au Point de Vue de la Constitution Médicale*, Paris, 1861, pp. 65–66.

16 At that time, many doctors used these kinds of therapy, though not necessarily with devices, as most relied on respiratory exercises (*gimnasia*). See, for example, the works of Jean-Charles T. Pravaz, *De l'emploi et du mode d'action de l'air comprimé dans le traitement des difformités du thorax*, Lyon, 1863, p. 22.

17 Vergara Lope, *Refutación teórica y experimental…*, op. cit. (n. 15), p. 16. Vergara's work is crisscrossed with references to sphygmographs and other instruments for analyzing

He felt that the flaw in the anoxihemia thesis arose from French pretensions to explain breathing on Mexico's high plateau from laboratories located in Paris and using sphygmographs on French patients accustomed to life on the plains. Vergara's explanation of this phenomenon relied on devices (sphygmographs, cardiographs) adapted to the mountain environment and experiments on people accustomed to living in those conditions. He argued that his cardiographic and sphygmographic records proved that Mexicans had developed physiological and anatomical modifications that allowed them to breathe quite adequately: a wider thorax, more frequent respirations and pulsations, greater respiratory capacity, and more red blood cells, a condition he called "hyperglobulia".[18] He believed that "the mixing of races [and] the climatic and even social conditions" modified "the organism to a greater or lesser degree", ultimately differentiating it "from other races".[19] Such modifications were not random, but adaptive variations proportional to the higher elevation and decreased atmospheric pressure. Vergara found, for example, that Mexicans made more respirations per minute than the French and so absorbed a similar amount of oxygen. Thus, he argued, the physiological differences between the two populations could be translated into physiological and, by implication, political norms.[20]

In the mountains, those instruments showed 'character'; that is, plasticity or mutability. Using the same European devices as Jourdanet, Vergara found that in spite of—or because of—these physio-anthropological differences, high-altitude dwellers in Mexico had as much oxygen in their blood as

cardio-respiratory functions. See, for example, D. Vergara Lope, "Notas sobre la Tensión Sanguínea en México", *Anales del Instituto Médico Nacional* 2 (1896), pp. 110–113: 110–111; D. Vergara Lope, "Algunas Experiencias de Oxigenografía. Oxigenógrafo del Dr. Fréderícq modificado por el Dr. Vergara Lope de México", *Gaceta Médica de México* 58 (1927), pp. 291–299; and D. Vergara Lope, "Una Nueva e Importante Aplicación de la Orto-Radiografía", *Gaceta Médica de México* 5 (1910), pp. 174–185: 174–175.

18 D. Vergara Lope, "La hiperglobulia de las altitudes no es un fenómeno de hematopoiesis", *Gaceta Médica de México* 8 (1912), pp. 417–424: 418–421.

19 D. Vergara Lope, "Biología. Algunas palabras acerca de la importancia de los estudios biológicos y antropométricos. Memoria Reglamentaria presentada ante la Academia Nacional de Medicina", *Gaceta Médica de México* 5 (1910), pp. 8–14: 9. We have to note that Vergara's studies stand on the transformation of human and animal races but he was critical of hypothesis of natural selection. He attempted to bring together the "best" of Lamarck and Darwin in order to prove that the environment influences the population just as much as selection and inheritance.

20 Homi Bhabha, *El lugar de la cultura*, Buenos Aires, 2002, pp. 48 and 59; Bernard S. Cohn, *Colonialism and Its Forms of Knowledge: The British in India*, Princeton, 1996.

sea-level dwellers in France. This meant that Mexicans were not abnormal or oxygen-deprived when compared to Frenchmen, a conclusion that led him to question whether those physiological contrasts were being employed by scientists to explain or justify the political and economic hierarchy of the two groups.[21]

At a meeting of the Mexican Academy of Medicine in 1907, Vergara presented a cardiograph, which he had designed and built for the physiology laboratory at the National School of Medicine. It was an almost exact replica of Charles Verdin's universal cardiograph, which was a copy of Marey's horizontal model preferred by experimentalists. Vergara held that Verdin's modified version "had certain defects", which he proposed to improve by modifying some parts, but without changing its form. In effect, Vergara himself observed that the cardiograph at the National Institute of Medicine seemed to be a copy of a French instrument. The modifications that he proposed were functional and focused on the central plate. He adapted a diaphragm to which the animal's heart would be subjected without it pressing on the thorax. He also modified the apparatus' support such that the animals were more firmly contacted. Aside from the cardiograph, Vergara and Juan Peralta, the Institute's mechanic who devised the new cardiograph, re-designed seven other French instruments. All those artefacts, modelled on the basis of devices that already existed, were unique creations that never found their way out of the Institute's Laboratory for Experimental Physiology, and thus never became commodities. Their value did not come from the market. In Vergara's words, the cardiograph "has the merit of being a national apparatus [which] teaches us that it is now completely feasible in our country to manufacture such delicate and precise apparatuses".[22]

His goal was to strive to increase the effectiveness of such devices for the study of the physiology of Mexicans, and held that European instruments and theories of high-altitude respiration had been developed by physiologists who were "as pernicious for the human species as the most bloodthirsty conquerors".[23] The heightened political sensibility of the time gave meaning

21 Vergara Lope, *Refutación teórica y experimental...*, op. cit. (n. 15), pp. 22–23.

22 D. Vergara Lope, "Cardiógrafo universal", Mexico, *Gaceta Médica de México* 3 (1907), pp. 109–114: 111.

23 Daniel Vergara Lope and Alfonso L. Herrera, "Un caso de anemia de los mineros y tuberculosis incipiente curado con baños de aire enrarecido", *Memorias de la Sociedad Científica Antonio Alzate* 10 (1896–1897), pp. 169–182: 174.

and value to instrumental practices structured around a disquieting, 'racist-nationalist' question: was the Mexican race normal and civilized?

Vergara's research and other projects carried out at the National Institute of Medicine's laboratories depended greatly on a market that involved European instrument-makers and Mexican officials from the Ministry of Development, specialists in locating devices, negotiating their purchase, and shipping them home. Though this trade in research equipment satisfied the needs of hospitals and medical researchers, in the long-term we may well ask whether this discouraged manufacturing in Mexico, where medical and scientific instrument makers were virtually non-existent. But what is certain is that at the beginning of the twentieth century, the commercial dominance of foreign instruments became consolidated, so the question is: why was instrument manufacturing not supported in Mexico, especially for devices like Vergara's cardiograph that were adapted for use at high altitudes and designed to reflect the respiratory and political normalcy of Mexicans?

Several possible explanations can be offered. One could be the tendency among Mexican doctors to espouse European theories and, by extension, their instruments. It is clear that Mexican doctors became familiar with European devices by reading the works of Frenchmen like Théophile Laennec, Bernard and Marey.[24] In addition to this theoretical influence we could also mention the lack of means for producing precision instruments. But this thesis—widely-held by historians—may be debatable. The trans-Atlantic journey of such instruments was long, slow and costly, and they were often found to be broken or incomplete upon arrival and required repair or replacement parts. Moreover, despite their contact with producers, practicing physicians had to follow cryptic written instructions from catalogues that made it impossible to follow step-by-step procedures. While there is no one sole answer to this question, if we approach it from the point of view of consumption we must ask, what local practices and policies were fulfilled or satisfied through the consumption of instruments from the European markets?

24 Purchases of instruments by the ENM were often inspired by the work of French and German physicians, such as Theophile R. Leannec, *Traité de l'Auscultation Médiate et des Maladies des Poumons et du Coeur*, 2 vols., 2nd Ed., Paris, 1826; Claude Bernard, *Leçons de physiologie opératoire*, Paris, 1877; J. E. Marey, *Physiologie expérimentale. Travaux du laboratoire de M. Marey*, Paris, 1875–1876, 4 vols.; and J. E. Marey, *La méthode Graphique dans les Sciences Expérimentales*, Paris, 1878, among others. See: "Catalogo de los utensilios y aparatos para las cátedras de fisiología y terapéutica", 1876: AHFM-AFEMyA, file 601.

Journeys through Political Inventories

As the historian Stephen Haber recognizes, Mexico began importing technology from Europe in the nineteenth century because it had no engineering or manufacturing industries, a fact he attributes to a lack of "indispensable engineering knowledge" that obliged the young nation to consume technology from the "advanced countries".[25] In his view, the absence of knowledge and ingenuity plus economic poverty explains the lack of artefacts and instruments. However, the way in which Mexican physicians adapted and reinterpreted those instruments casts doubt upon the validity of this explanation, especially if we also accept, with Simon Schaffer, that the repair and modification of these instruments entailed generating new knowledge and creating new scientific practices.[26] Of course, ingenuity and knowledge do not translate automatically or necessarily in engineering skills and industrial resources for instrument production. Upon reflection, what does become visible is that those instruments added value and offered political and cultural compensations to government functionaries and contemporary doctors. Beyond the prices of such instruments as foreign commodities, their importation strengthened the policies of the time that sought to create a nationalist medicine that would agglutinate the indigenous populations into one sole Mestizo (mixed race) nation.

Clearly, the scientific knowledge and practices of Mexican doctors were not isolated from the intense mobilization of people, artefacts, and scientific theories produced by colonial expansion.[27] My research into the local contexts where instruments were used shows that the movement of ideas and devices was neither one-way (Europe to America) nor a simple matter of imitation and diffusion. Rather, it required a network of political relations and knowledge that involved both areas. A review of purchase orders from Mexican medical institutions to European instrument-makers (especially French ones) shows that this trade existed thanks to a group of doctors and engineers hired by Mexico's Ministries of Development and Public Instruction / Fine Arts for their specialized knowledge, contacts, and familiarity with the trading networks where such devices circulated. Those physicians and engineers, as well as the officials

25 Stephen Haber, *Industria y subdesarrollo. La industrialización de México, 1890–1940*, Mexico, Alianza Editorial, 1992, pp. 46–47.

26 Simon Schaffer, "Easily Cracked: Scientific Instruments in States of Disrepair", *Isis 102* (2011), pp. 706–717.

27 Bourguet, Licoppe and Sibum, op. cit, (n. 3), p. 4, and R. MacLeod (ed.), "Introduction", *Osiris* 15 (2000), p. 10.

at the *Ministerio de Fomento*, knew instrument makers like *Maison Mathieu et Co., Maison Charles Verdin*, and *Charrière et Collin*.[28] The remarkable facet of their work was not how they facilitated this trade, but that they performed the functions of evaluating which instruments to acquire and choosing which research projects to support. In some cases, in fact, they were wealthy enough to set up commercial houses in Mexico City and trade in such instruments.[29]

An excellent example of the enterprising spirit of these agents was Carlos Pacheco, a professional military, intimate political ally of Porfirio Díaz, and *Secretario de Fomento* (Secretary of Development) from 1881 to 1890, who fought to create the National Institute of Medicine and zealously assure that its physiological and chemistry laboratories were well-equipped. The commercial activity of such officials fulfilled a political mission: supporting the task of adapting, modifying, and re-designing European instruments. In reality, no policy of stimulating instrument-making existed because those activities were controlled by the colonial technological empire. However, importing and re-designing some apparatus provided those with nationalist sentiments with a means to affirm difference, claim agency, and develop a creative capacity. It allowed Mexicans to reaffirm their normality in spite of European attitudes that regarded them as inferior.

Making this trade a reality required a network of political relations and agreements that enabled instrument cargos to be transported from Europe as objects of global commercial consumption. And that effort depended greatly upon the European manufacturers and their relations with doctors and

---

28 These three houses of medical instrument makers are those most often cited, but the *Charrière et Collin* perhaps made more transactions in Mexico; see, "Escuela Nacional de Medicina informa de los instrumentos llegados a Veracruz", 1873: AHFM-AFEMyA, file 534; "Catálogo de los utensilios y aparatos para las cátedras de fisiología y terapéutica", 1883; AHFM-AFEMyA, file, 501; and "Lista de los objetos pedidos a Europa en esta fecha", 1873: AHFM, AFEMyA, file 534. The *Charrière et Collin* trading house was founded by Jean Fréderic Charrière (1803–1876), who was trained as a *coutelier*. Upon his death, he was succeeded by his son, Jules, and his apprentices Robert and Adolphe Collin. On the *Maison Charrière*, see: *Industrie Maison Charrière-Collin, Catalogue Général Illustré d'instruments de Chirurgie*, Paris 1878, Paris, 1885, pp. 1–2; Audrey B. Davis, *Medicine and Its Technology. An Introduction to the History of Medical Instrumentation*, 2 vols., Westport, CT, 1981, pp. 20 and 22; and Alain Ségal, "Les Couteliers et fabricants parisiens d'Instruments Chirurgicaux. L'Essor de l'Instrumentation Chirurgicale au XIXeme Siècle", in Alain Pecker (ed.), *La médicine à Paris du XIIIème au XXème Siècle*, Paris, 1984, pp. 240–243.

29 Mexico was not the only country that experienced this, as European and North American physicians were similarly affected; see Davis, op. cit. (n. 28), p. 21.

functionaries in the Americas.[30] This political aspect of trade in European-made instruments emerges clearly in the case of the purchase of Legay's apparatus, among others, by the National Institute of Medicine. This strange and, it must be said, costly instrument (11,634.10 francs, or 2326.8 silver pesos)[31] consisted of a large chamber with a pump to compress air and a thermosyphon to heat or cool it. Though oxygen rarefaction was a common therapy for respiratory diseases, neither the apparatus nor its creator were well known in France or Mexico. Presumably, it was obtained through *Paul Rousseau et Cie.*, a Parisian medical instrument maker and seller whom doctors from the National Institute of Medicine have met at the 1889 Universal Exposition in Paris.

In any case, the device was obtained in 1891 for reasons of a strictly personal nature: worried about General Pacheco who suffered from pulmonary emphysema, Dr. Ignacio Altamirano decided to try a therapy using Legay's apparatus. When the transaction was consummated some ten months later it became just that. Rousseau recounted that the apparatus had not been purchased before, so it took almost ten months to complete its sale to Mexico, after negotiating the total number of parts and the final price. Apparently, Rousseau failed to inform the Mexican doctors that the catalogue price did not include the pump, an omission that led to heated arguments among them and members of the Ministry of Development, and a delay in completing payment for the device. Once it arrived, it took almost a month to put it together and make it work, as the doctors never received assembly instructions. When they finally got it to work, General Pacheco had been dead for nearly a month![32]

30 See how Schaffer interprets this idea of cargo, based on Greg Dening: S. Schaffer, "Instruments as Cargo in the China Trade", *History of Science* 44 (2006), pp. 217–246 and Greg Dening, *Performances*, Chicago, 1996.

31 A calculation based on receipts from Rousseau; see: "Letter from the Minister of Development to P. Rousseau", México, 7 October 1891: AGN, IMN, Box 125, file 1, fols. 176–184.

32 "Concerning the pump, [this Ministry] has information that it has arrived in Veracuz, but not in the capital". The apparatus later reached this City, "and [we] shall proceed with its examination and installation but, as feared, the aforesaid apparatus arrived too late, as General Carlos Pacheco died on the 15th of last month, so it was not possible to use [it] to cure him, though that was the main objective in ordering it": ibid., fols. 138–138v. From his side, Rousseau said that the doctors on the other side of the Atlantic had not understood that the price of the artifact did not include the accessories (thermosyphon, compression bells) and that the Mexicans claimed that the packing and, above all, the assembly instructions for the artifact were not clear. An analysis of the images of any instrument

In 1893, a few years after its assembly and start-up, Legay's apparatus was re-created by doctors at the Institute. No longer a device obtained to cure the Minister, it was converted into a medical instrument that acquired a certain commercial value when Vergara Lope and his colleague, Alfonso L. Herrera, created their "Aerotherapy Cabinet" that offered respiratory patients therapy sessions using compressed air at a cost of two pesos.[33] Despite their difficulties and misunderstandings with Rousseau, Mexican doctors continued to purchase bizarre and expensive instruments; unique specimens that were used in experimentation until they were commercialized, as in this case, to cure by using the benefits of artificial pressure to improve respiration.

In Mexico, such instruments were reproduced not only as emblems of precision but also to address differences among bodies in political terms; i.e., the question of racial differences. Noteworthy here is the fact that during their travels, the physicians who practiced medicine from laboratories equipped with precision instruments played a key role in mobilizing stereotypes of authorized and universal (*qua* European) knowledge and in conforming/confirming new simplifications or stereotypes of non-Europeans. Though on the one hand they affirmed the normality of Mexicans, their findings (sphygmograph readings and other physiological measurements) also raised up the Mestizo (those of mixed race) as the new, local model of normality established by instruments. If people thought that Mexicans were abnormal when compared to Europeans, then the Indian, the poor, and criminals who deviated from the ideal normality of the Mestizo, became seen as abnormal by high-altitude, middle-to-upper class Mexicans (and worse by Europeans) and not included in the nationalist agenda of the period.

Support for importing instruments was one expression of a policy that favoured trade and knowledge. Though isolated from profit-making activities, doctors were integrated into import operations as creators through a logic that was almost artistic in nature; one based on artisanal reconstructions that had no technical pretensions or commercial reproducibility, but that endowed the instruments with a certain political value. The trade route between America and Europe was not predicated on profit-making but, rather, on political interests that sought to re-create and confirm—through artefacts and theories—great national questions: the normal political authority of Mexicans living

catalogue from the period makes it easy to appreciate the difficulties that anyone at a great distance would have faced in assembling and starting up such artifacts.

33 Ignacio Altamirano, "Letter to the Minister of Development", México, 1901: AGN, IPyBA, IMN, box 126, file 1, fols. 112 and 114–114v.

at high elevations. The ways in which the instruments were used shows the extent to which knowledge is not only repetition (generalities, laws) but also events and local adaptation. The life of instruments is debated in the tension they reproduce like an echo, in the inequalities and hierarchies among which knowledge is played out: between the economic interests of commercial industry and the political interests of a local medical elite that sought to confirm itself, through difference, as modern and civilized.

General Index

www.ingramcontent.com/pod-product-compliance
Lightning Source LLC
LaVergne TN
LVHW010545300926
849162LV00002B/235

9789004324923